A MATHEMATICS LEADER'S GUIDE TO

LESSON STUDY IN PRACTICE

A MATHEMATICS LEADER'S GUIDE TO
LESSON STUDY IN PRACTICE

Jane Gorman
June Mark
Johannah Nikula

Education Development Center, Inc.

Newton, MA

HEINEMANN
Portsmouth, NH

Heinemann
361 Hanover Street
Portsmouth, NH 03801–3912
www.heinemann.com

Offices and agents throughout the world

This work is supported by the National Science Foundation under Grant Numbers ESI-0554527 and ESI-0138814. Any opinions, findings, conclusions, or recommendations expressed here are those of the authors and do not necessarily reflect the views of the National Science Foundation.

Library of Congress Cataloging-in-Publication Data
Gorman, Jane.
    A mathematics leader's guide to lesson study in practice / Jane Gorman, June Mark, and Johannah Nikula.
        p. cm.
    Includes bibliographical references.
    ISBN-13: 978-0-325-02799-9
    ISBN-10: 0-325-02799-4
    1. Mathematics—Study and teaching. I. Mark, June. II. Nikula, Johannah. III. Title.
QA11.G675 2010
510.71—dc22   2009046255

Editor: Victoria Merecki
Production: Lynne Costa
Cover and interior designs: Shawn Girsberger
Typesetter: Shawn Girsberger
Manufacturing: Valerie Cooper

Printed in the United States of America on acid-free paper
14   13   12   11   10   ML   1   2   3   4   5

Contents

Acknowledgments

A *Mathematics Leader's Guide to Lesson Study in Practice*, similar to the companion materials (*Lesson Study in Practice: A Mathematics Staff Development Course*), is the result of a decade of experience working together with many teachers and students to improve mathematics teaching and learning using lesson study. We are particularly grateful for the teams of teachers who participated in our Lesson Study Communities in Secondary Mathematics project, which provided the basis for the ideas and examples that appear in this *Leader's Guide*. That project included teachers from the following Massachusetts schools:

Ahern Middle School, Foxborough
Andover High School
Bartlett Middle School, Lowell
Brookline High School
Carver High School
Danvers High School
Dartmouth High School
Everett High School
Lexington High School
MATCH School, Boston
Newton South High School

Norfolk County Agricultural High School, Walpole
Old Rochester Regional Junior and Senior High Schools, Mattapoisett
Scituate High School and Gates Intermediate School, Scituate
Uxbridge High School and Whitin Middle School, Uxbridge
Wareham Middle and High Schools
Watertown Middle and High Schools

The teachers' experiences learning about lesson study, in their efforts to make their teaching of mathematics more effective for their students, were invaluable to our work and their thoughtfulness and willingness to share their learning with others have been a tremendous contribution. These teachers demonstrated through their work a powerful commitment to their colleagues, to their students, and to their own learning. We are very thankful for the support of school and district administrators in these schools for making teachers' participation possible. The quotes that appear throughout this book capture the words and voices of the mathematics teachers from these teams who have walked alongside us in our learning about lesson study. We particularly want to thank the authors of the team stories that make up Part IV for sharing their own experiences as teachers, coaches, and administrators so that others can envision lesson study in practice.

We also want to thank the teams and facilitators from the following districts, who pilot-tested the *Lesson Study in Practice: A Mathematics Staff Development Course* (the *Course*) or worked with us as part of other district or state initiatives, therefore providing us further insight and ideas: Baltimore City Schools, MD; Chelsea Public Schools, MA; New York City Public Schools, NY; Springfield Public Schools, MA; Watertown Public Schools, MA; West Virginia STEM Center; Westport Public Schools, CT; Winchendon Public Schools, MA; and Winsor Public Schools, CT.

Supporting the work of these many school teams was a group of coaches. In addition to our own work as coaches, the following people contributed: Joan Bryant, Hillary Dockser Burns, Loretta Heuer, Euthecia Hancewicz, Joseph T. Leverich, Diana Metsisto, and Deborah Spencer. As a group, we met periodically to discuss issues related to coaching lesson study in mathematics, and those discussions shaped much of our thinking about lesson study leadership.

The drafting and reviewing of chapters by Joseph T. Leverich, Euthecia Hancewicz, and Hillary Dockser Burns were critical to the completion of this *Leader's Guide*. In addition, throughout our work in lesson study we have benefited from the support of our advisors Thomas Judson, Catherine Lewis, Deborah Schifter, Akihiko Takahashi, Phyllis Tam, and Tad Watanabe. We also wish to recognize the powerful ideas contributed by Brian Lord, who led our project research. These advisors have taught us about lesson study, teaching, mathematics, and colleagueship. The knowledge they so generously shared with us is embedded throughout the chapters of this book.

We also wish to thank our evaluator, Joan Karp at the Program and Evaluation Group at Lesley University, for providing valuable feedback and advice. We gratefully acknowledge the support of the National Science Foundation for funding the projects (ESI-0138814 and ESI-0554527) that made this work possible.

PART I

Introduction to Lesson Study Practice

About This Guide

esson study is a structured approach to studying, developing, and improving lessons. It is a cycle of inquiry about student learning, conducted for the purposes of teacher learning and instructional improvement. Sometimes taking place among colleagues at a school, and sometimes in cross-district teams or conference settings, lesson study provides educators the opportunity to talk in depth about how students learn mathematics, using live classroom lessons as the basis for discussion and learning. Emerging in the United States in the late 1990s, lesson study is modeled on the Japanese practice of *jugyo kenkyu*,[1] and is now conducted by teachers in many countries, worldwide. Lesson study is classroom-based research about how students think and learn, and is conducted by small teams of teachers who share common goals. Each team's intensive study of content, teaching methods, and student learning within one research lesson is a contribution to a larger, ongoing process of professional knowledge building, both within and across schools. The cycle of lesson study (see Figure I–1) provides a powerful opportunity for the teachers on the lesson study team to learn, and to improve their own students' learning. Their research lesson offers equally important opportunities for many other educators to learn—by participating in the lesson observation and post-lesson discussion, or by reading the team's written research lesson reports and studying the research lesson instructional plan.

The cycle illustration summarizes what the steps in lesson study are, and by itself has provided many teachers with a model for beginning lesson study. It was with little more than this image that in 2001 we did our first cycle of lesson study with a team of high school mathematics teachers. This simplicity was enough to get us started! The first cycle was powerful, though very rough around the edges. It eventually led to a sustained district initiative, and also to opportunities for us to work with many other teams, over many cycles, through the

Focus, Set Goals, and Research the Topic

Develop the Research Lesson

Teach, Observe, and Discuss the Research Lesson

Reflect, Consolidate, and Share Learning

Figure I–1

---

1 Japanese teachers have used lesson study since the nineteenth century to implement new forms of instruction and build teacher knowledge. The term *lesson study* (*jugyo kenkyu*) refers to the process used to develop, observe, discuss, and reflect on a research lesson or study lesson (*kenkyu jugyo*). See the lesson study glossary at http://hrd.apecwiki.org/index.php/Lesson_Study for more information about this and other lesson study terminology.

NSF-funded Lesson Study Communities in Secondary Mathematics project.[2] Through the insights teachers have shared with us and through our own gradual development as lesson study coaches, we have become more aware of what new teams need to get started, and what it takes for them to go forward to build and sustain a strong, lesson study practice.

Most teams probably begin, as we did, with following the steps of the cycle. Gradually over time, with much trial and error involved, teams can move to a deeper level of practice by building expertise and leadership within the team and bringing in knowledge and support from outside the team. For both teams and leaders, this shift involves new skills and ways of working with colleagues, and also development of a larger vision of the work. The purpose of this guide is to help teams and leaders make this transition to deeper lesson study practice and, ultimately, to build professional knowledge and improve students' learning of mathematics. To do this, we've translated the help we have been providing teams and the insights we have gained over the years into practical advice for the reader. This guide is designed to help you to envision, understand, and achieve a deeper level of lesson study work.

Content of the Guide

The content of *A Mathematics Leader's Guide to Lesson Study in Practice* (referred to as the *Leader's Guide*) has emerged from our experience of coaching lesson study teams through multiple cycles and from the many informal conversations we have had with teachers and leaders. In writing the *Leader's Guide,* we have considered: What questions do teachers and leaders ask us about lesson study? What parts of the process have been most difficult for teams to understand and do well? What stumbling blocks and barriers have we seen over and over? Which stories about teams and their lesson study work have we told over and over, because they have been so interesting and helpful to other teams?

It is our experience that teams thrive and grow when they pay attention to three main realms of lesson study learning:

- Developing a vision of the larger goals, themes, and principles that underlie lesson study, and the ability to apply these to work on their research lessons

- Developing skill, experience, and understanding in the specific tasks and activities that make up the lesson study cycle

---

2   The Lesson Study Communities in Secondary Mathematics project was supported by the National Science Foundation under Grant Number ESI-0138814. Any opinions, findings, conclusions, or recommendations expressed in this *Leader's Guide* are those of the authors and do not necessarily reflect the views of the National Science Foundation.

- Building understanding of the multiple connections their work has to the work of their school, district, and the larger profession, and activating these connections in support of their work

The *Leader's Guide* content is organized with these three realms in mind to best support teams and leaders in deepening their lesson study practice over time.

Part I: Introduction to Lesson Study Practice

The first part of the *Leader's Guide* introduces and summarizes the big ideas or themes that underlie lesson study work and that will be developed throughout the rest of the guide.

Part II: Deepening Lesson Study Practice Across the Cycle

For each phase[3] of the lesson study cycle, the *Leader's Guide* contains a brief phase overview and topical chapters on deepening the work of individual lesson study teams during that phase. Each chapter is focused on a specific lesson study process and contains examples drawn from our work with teams. A timeline of a typical lesson study cycle is provided so that teams and leaders can easily locate information that supports their work and pushes their thinking as they move through the cycle.

Part III: Building Sustainability and Connecting to the Wider Profession

Part III focuses on how the work of the individual team connects to the larger professional context and develops sustainability. Chapters discuss issues such as building support for lesson study in the school, hosting public research lessons, the role of administrators, and connecting with outside expertise.

Part IV: Envisioning Lesson Study Implementation and Practice

In addition to weaving examples of lesson study teams' work into the chapters, the *Leader's Guide* also contains a number of narratives, each describing the work of one team across one or more cycles and sharing something of the history of the team— how they developed their lesson, what their big goals and challenges were, and so on. These chapters offer the reader firsthand experiences of how these teams began and continued lesson study practice in their schools.

Resources

Resources are suggested throughout the *Leader's Guide*, and an annotated bibliography of lesson study and mathematics resources is provided in the Resources Appendix.

---

3   There are no formal divisions in the lesson study cycle, but most teams think of their work as falling into the four phases or stages pictured in Figure I–1: focus, set goals, research the topic; develop a research lesson; teach, observe, and discuss the lesson; and reflect, consolidate, and share learning.

▋ Using the *Leader's Guide*

This *Leader's Guide* is written for *teachers* who are doing lesson study and who want to learn more about it or who are playing a lead role in organizing a team, for *coaches* and other *math leaders* who support lesson study teams, and for *school administrators* who have lesson study going on in their schools or want to introduce it. The *Leader's Guide* is also a companion resource for *facilitators* of *Lesson Study in Practice: A Mathematics Staff Development Course.*[4] For all readers, the *Leader's Guide* describes the features of strong lesson study practice, and discusses how teachers and leaders can work together over time to develop this strong practice and build sustainable lesson study programs in their schools. The key words here are *over time* and *develop*. One's lesson study practice should continue to grow and deepen across one's entire career.

The *Leader's Guide* is intended to be used as a resource that teams and leaders will turn to many times, for many purposes, in reflecting upon and supporting this continued growth. As you read a chapter, don't approach it as a user's manual with instructions that must be followed fully. Think of it as food for thought about the topic, a set of ideas that other lesson study teams and leaders are sharing with you, to help you deepen your learning about mathematics and teaching.

Some ways teams and leaders might use the *Leader's Guide* include:

- *To introduce people to lesson study*: Reading a team story, the phase overviews, or brief selections from the chapters could bring lesson study to life for teachers or administrators, perhaps in the context of a brief introductory workshop.

- *To enrich team practice*: Many teams may have started their work with a different model of lesson study, or with a minimal understanding of what it entails. For these teams, reading essays in the *Leader's Guide* might suggest new ways of working, or new steps in the process that they have not tried before.

- *To address challenges your team is facing*: Reading one chapter about something your team is struggling with could spark a team discussion or provide ideas for improving that aspect of your lesson study work.

- *To build lesson study leadership capacity*: Teachers and coaches (from experienced lesson study teams or new to lesson study) often take on a leadership role in the school around lesson study, mentoring new mathematics teams or teaching colleagues in other subject areas about lesson study. The *Leader's Guide* provides a strong resource for these leaders, taking them beyond the level of having done a cycle or two, to thinking about what makes strong practice and what the role of leaders is in supporting this work.

---

4   The resource that is referenced here, and later in this introduction, is Gorman, J. et al. 2010.

- *To support district implementation and planning*: District mathematics leaders and administrators may find the *Leader's Guide* to be a useful resource in planning system support and implementation of lesson study in multiple schools.

Teacher Use of the *Leader's Guide*

Teachers on a lesson study team may want to use selected readings from the *Leader's Guide* to support reflection and improvement around specific areas of the team's lesson study practice or to build understanding of the big themes that drive the work. The *Leader's Guide* contains many examples of work from the teams with which we have worked. These examples can give concrete images of the big ideas of lesson study, help you get a stronger sense of whether your work is moving in productive directions, and answer specific questions, like:

- What is meant by *researching the topic* and what questions should we be asking when we do it?

- How do teams with teachers from multiple grade levels focus their work and decide what grade lesson to teach?

- If we invite teachers from outside the team to our lesson, how should we prepare them for the observation?

- What is the purpose of using the multicolumn lesson plan and how does creating the plan deepen team learning?

Team Leader and Lesson Study Coach Use of the *Leader's Guide*

Many teams have ongoing mathematical and pedagogical support from a teacher leader, a mathematic specialist or coach, a department head, or possibly someone from outside the school (e.g., a university professor, professional development leader, or lesson study specialist). As a coach or team leader, you will find help within the *Leader's Guide* for building your general understanding of lesson study and guidance related to specific parts of the lesson study process. Questions we have faced in coaching teams, or that we know are challenging for many team leaders, are treated in the *Leader's Guide*. For example,

- What does *deepening lesson study practice* mean?

- What are the challenges a team might face in each phase of the cycle? For example, what challenges might the leader anticipate facing during the lesson development process?

- How does a team improve the lesson observation to increase team learning?

- How can a leader support teachers on the team in being willing to take the risk of teaching a lesson with many observers?

- What is the leader's role at the post-lesson discussion? Are there protocols for the moderator to follow?

- How does leading lesson study differ in approach or goals from the role as mathematics specialist or coach?

- How does the team develop internal leadership and collaborative norms?

School Administrator Use of the *Leader's Guide*

The principal, assistant principal, and district mathematics leadership staff have powerful reasons to encourage the development of a lesson study program in their schools. Yet, if lesson study is just being introduced, or has started up with a few volunteer teachers, it may not be clear what the best ways are to support individual teams or to implement lesson study at a schoolwide level. As an administrator, you will find help throughout the *Leader's Guide* on questions such as:

- How have other districts used lesson study to support their mathematics program?

- What is an appropriate role for the principal in the lesson observation and post-lesson discussion?

- How do the lesson study team's goals relate to schoolwide goals for students?

- How do lesson study teams utilize student assessment data in setting goals or choosing a content focus?

- What are the practical supports such as time, money, and coaching, needed by lesson study teams?

School administrators may wish to introduce lesson study to district math coaches or lead teachers as a preliminary step in schoolwide implementation. One model for this training would be for coaches to participate in the *Lesson Study in Practice: A Mathematics Staff Development Course*, periodically stepping aside from the regular activities in the course for leadership discussions based on readings from this *Leader's Guide*. This would provide a personal experience in *doing* lesson study, information about effective lesson study practice, and opportunities to reflect and consider leadership issues.

|||||||||||||||||||

Whether you are a teacher, coach, or school administrator, we hope that this *Leader's Guide* provides you with inspiration and ideas to initiate and deepen your own lesson study practice. Many teams encounter challenges early on in their work, and establishing lesson study as a new form of professional learning in a school can require changes in thinking and organization that take time and leadership. However, as teams and leaders continue to focus on improving their practice they are rewarded with insights about their students, about mathematical understanding and how it develops, about their colleagues, and about themselves as teachers and learners.

It is important to remember that lesson study is about learning—you and your colleagues' learning and your students' learning—and to take advantage of the many opportunities for that learning while engaged in lesson study work.

CHAPTER 1

Developing Lesson Study Practice

▍ Big Ideas in Lesson Study

The Forest and the Trees

As a lesson study team member or leader working with a team, there is nothing worse than being lost in a series of detailed decisions about the lesson, or focused on the mechanics of the lesson study process without understanding why they matter or how they might eventually contribute to better teaching and learning at your school. In the beginning, teams do focus a lot of energy on learning the steps of the process. Whether in the first cycle or the tenth, however, it is essential to have a sense of the bigger picture, to have a few beacons to follow. Understanding the big ideas that drive the work, and weaving them into the fabric of your team's practice, is probably the most powerful strategy you can have for taking lesson study to a deeper level. These are the big ideas that have been the strongest beacons for our teams, and for us as leaders.[1] These ideas are briefly described here, and will be seen as recurring themes throughout this book:

- Lesson study as research

- Focus on mathematics

- Focus on students

- Professional community

Lesson Study as Research

Who best to research the intricate details of teaching daily lessons than teachers themselves?[2]

---

1   These themes are consistent with research from other U.S. lesson study sites pointing to essential elements in lesson study (Lewis 2002b). What are the essential elements of lesson study? *The CSP Connection,* *2*(6), 1, 4), pathways that connect lesson study to instructional improvement (Lewis, Perry, and Hurd 2004). A deeper look at lesson study. *Educational Leadership, 61*(5), 18–23), and important stances teachers must adopt (Fernandez, Cannon, and Chokshi 2003). A U.S.-Japan lesson study collaborative reveals critical lenses for examining practice. *Teaching and Teacher Education*, 19, 171-185)
2   This is the first of many places in this book where a teacher who has participated in lesson study is quoted. Unless otherwise footnoted, all teacher quotes that appear in the *Leader's Guide* are reflections made by participants in the Lesson Study Communities in Secondary Mathematics project, NSF Project ESI 0138814, or by participants in the Resources for Supporting Lesson Study in Mathematics project, NSF Project ESI 0554527. Both are projects at Education Development Center, Inc. (EDC), Newton, MA.

Lesson study is a cycle of research or inquiry. The team develops a key question or hypothesis to be investigated through the research lesson and uses the observation as a strategic opportunity to capture data about student understanding. The fundamental assumption is that classroom instruction and learning can be improved over time through this experimental process. The team uses the lesson plan as an analytic tool to reflect on teaching and the lesson itself as data for improving instruction. As teams deepen their work, this vision of lesson study as research strengthens and the idea that lesson study is solely about planning model lessons recedes. The team's research lesson plans begin to reflect the team's research questions. Team decisions and conclusions are based on evidence and the team is eager to report on and make sense of data gathered through the observation rather than jumping right into revising the lesson plan. The goal is for the team to learn from their lesson study and see it as contributing to an ongoing research and development system for teaching—a system in which teachers are primary contributors.

Focus on Mathematics

> *Until lesson study we never discussed the value of the content being taught. We never really discussed how students learn a particular concept, what to expect for outcomes, and how to adjust the lesson to meet student needs. We have discussed the different ways students learn (multiple intelligences), how the brain works, how to differentiate an inclusion class. Never have those discussions involved a discussion of how to develop problem-solving techniques, how to develop a particular concept, how much depth should be expected for a particular concept. Lesson study made us focus on the content of the lesson—what exactly is the content goal for the lesson, and then how do we involve students in the learning of that concept.*

Lesson study is research about how students learn. As such, it places the highest priority on the discussion of how students come to understand particular concepts and on deepening teachers' knowledge of this mathematics.[3] Often, this mathematical focus comes into play naturally as an expected part of the process—for example, when the lesson content understanding goals are drafted or when the teachers share expertise about how the topic is taught in different grades. But there are also many moments in lesson study when the mathematics comes into the foreground because one good question gets asked. For example, in studying your textbook treatment of the lesson topic, someone might ask, "Why do these topics appear in this particular order in our

---

3   Lesson study is not limited to mathematics. While this book is aimed specifically at mathematics teachers and leaders, it is a resource about deepening lesson study that would apply in any content area. Teachers from other content areas may occasionally want to substitute or imagine examples from their subject area.

book?" In anticipating how students might solve the lesson problems, someone might ask, "Has anyone seen alternate algorithms being used on this kind of problem? What are they? Why do they work?" The mathematical focus comes into play when teachers are doing mathematics together before designing the lesson—including challenging mathematical extensions that cover a topic in a wide range of grade levels. It is present when teachers are trying to understand the mathematical context of the lesson, how students develop understanding of the topic across the grades, and understanding goals that reflect this content. It demands attending to students' mathematical thinking and strengthening teachers' mathematical understanding. And it means that decisions about the instructional design of a lesson are considered for their mathematical merits and for their potential impact on understanding.

Focus on Students

Most valuable to me so far [in our lesson study] has been the lesson I observed. I think I had forgotten that I should listen to the kids talk about math in math class. Sometimes, we get on the fast train to finish what we have to cover. I lose sight of what I want to hear from the kids. Listening to the students was an eye opener. They do care about being problem solvers.

Have you ever watched a classroom video and realized that the camera was pointed entirely at the teacher with no attention to student voices, mathematical work, interactions, and learning? In lesson study, research lessons are observed live by the whole team, specifically so that valuable information about students' learning *can* be revealed. This opportunity to collect detailed data on students' mathematical discussions is quite rare in most teachers' lives.

Focusing on students is extremely important during the observation, but is actually a powerful force for teacher learning *throughout the cycle*. Teams set student-centered goals related to the most pressing learning needs of their own students. They strive to understand the mathematics through a student lens. Understanding goals are framed to build on students' likely prior understandings and to connect to topics students will encounter in the future. The team explores how students think about the lesson content. Lessons are structured to encourage active student problem solving and to reveal student thinking so that the team can learn *how* students build understanding as well as *what* students are, or are not, understanding. At the post-lesson discussion, data on students' problem-solving approaches, errors, misconceptions, and varied ways of thinking about the mathematics are shared, and help the team learn what effective instruction for this topic might involve. Data on how the students feel or act is also noted so that the team can discover ways to foster their broad student-learning goals in their daily lessons.

Professional Community

> *There is a collegiality that has been built and lesson study definitely provides sustenance for that. It creates a forum where if [a teacher has] a question, they can ask a colleague, and not feel like, "Oh, I should have known that."*

> *We feel more invested in the [lesson study] process, as we have learned a great deal from each other and feel supported by this camaraderie.*

> *We were more willing to take risks in the conversations we had regarding issues that came up during lesson development.*

Lesson study is a teacher-driven collaborative learning model. Teams of teachers work and learn together over an extended time, setting their own goals and developing internal leadership, creating the lesson and observing it together, and discussing and debating theories about student learning within the team and with educators from outside. The collaborative aspect of lesson study is extremely rewarding for teachers, in particular because most teachers have worked in schools where professional isolation is the norm. Teachers' comments reflect this appreciation.

The team is also part of an extended professional community as they invite outside expertise to contribute to their learning and share their findings professionally. This sense of community on all levels needs attention for the team to thrive. Trust has to be nourished for teachers to open their classroom to observers, put their ideas up for debate, or reveal gaps in their knowledge to colleagues. Developing these and other features of what Brian Lord has termed *critical colleagueship* (Lord 1994) is by no means automatic. As a lesson study team launches and develops its practice, attending to community building and to developing a new openness to public scrutiny and evidence-based approaches will provide many practical benefits—such as having the team function well, building leadership, and reinforcing stronger collegiality. Research by Perry and Lewis (2008) suggests that the development of professional community is one of the key conditions of sustainable lesson study practice.[4] They also report that professional community is one of three mechanisms through which lesson study leads to instructional improvement.

▌Beginning Lesson Study—the First Cycle

Although the big ideas of lesson study as research, a focus on mathematics, a focus on students, and professional community develop over time in a team's lesson study

---

4   "Lesson study enables teachers to strengthen professional community, and to build the norms and tools needed for instructional improvement . . . increased motivation and capacity to improve instruction, norms that emphasize inquiry and continual improvement, sense of mutual accountability to provide high quality instruction, shared long term goals for students, and shared language, processes, and frameworks for analyzing instruction" (Perry and Lewis 2008).

practice, many benefits of lesson study emerge even during the team's first cycle of lesson study. Looking at the basic structure of the lesson study process (teachers gather, set common goals for students, study texts and other resources to deepen their understanding of content, then develop and jointly observe and discuss a research lesson), its benefits are easy to imagine. To appreciate some of the benefits that surface even as a team works on their first research lesson, we can listen to the voices of a few teachers who are reflecting on their first cycle of lesson study.

Thinking Deeply About Lessons

> For my team, it has been an eye-opening experience. It made us realize that there is so much more that we can do to improve our teaching. Participating in lesson study has made us think. Even though we always thought through lessons before, we now think about them in a different way. . . . We really took time to discuss what we wanted our kids to get out of the lesson and explore the possible difficulties they may have with the concept . . . we did not have to accept our first idea for the lesson as the one we would use. We had time to consider many ideas. We also had time to make a more creative, interactive, thought-provoking lesson than we normally would.

Developing Professional Community

> In our first cycle of lesson study, the process has brought the department members who are involved together as a team. We are willing, to a much greater degree, to work together, ask each other questions, and help each other out. The lonely feeling many of us have when we teach day in and day out with very little interaction with our colleagues is diminished. Now that I am closer to my departmental colleagues I think of them more often, and want to help them if I can, and I am willing to have them help me.

Seeing Students from a New Perspective

> I found that an integral part of the process was the classroom observations. When we planned the lessons, we were asking students to think in ways that they were not accustomed to. So, we really had no idea what ideas and processes that they would come up with. Therefore, the classroom observations were critical in the development of our lessons. Often when we teach a class, we don't have time to listen to what students are thinking individually or together in small groups. This is because we are too busy moving around the room, helping our students to focus on the tasks at hand. The observations during our [research] lessons provided unique opportunities to sit and listen to what students think. Personally, this gives me great insight not only into how students were thinking about the specific concepts, but also how they work together in groups in general.

All of these teachers' comments are not unusual. In working with lesson study teams over a period of many years, we have learned that the first lesson study cycle a team experiences is often powerful and rewarding in many ways. But we have also learned that for most teams, doing lesson study for the first time is an exploration of very new professional territory. The way the whole lesson research process unfolds, as well as the specific tasks, conversations, and ways of collaborating, are new for most teachers, as is illustrated by the quotes that follow.

Needing New Skills

When we first began this process last September, I didn't give much thought to the skills that I would need to possess in order to implement lesson study in our school. Yet, now I have a newfound appreciation for this part of the process. . . . In the fall cycle we did not do a very good job at all of observing classes. In the spring, we identified this as an issue and focused on making good and effective classroom observations. However, we failed to communicate effectively to the other teachers involved what they should be doing as observers. Thus, our post-lesson discussion wasn't nearly as effective as it could have been.

Lacking the Overall Perspective

It is the end of the school year and I feel that [as an individual] I am beginning to acquire a much stronger grasp of the ideas behind lesson study. . . . As a department, I feel that we are still searching for a clearer perspective. Our initial lesson study work was coopera-tive but lacked the understanding of how to choose a lesson and what would be the focus of what we wanted to achieve. I feel that we were more concerned with what to teach rather than how to teach it. I feel that next year's lesson study will be much improved in focus and execution.

Shifting Long-held Viewpoints

I still find it challenging to place the focus on students rather than on teaching. This is a very hard distinction for teachers in my team to make. We see the lesson from our own point of view, and shifting to the students' point of view takes a concerted effort. It was easier to focus on students in the post-lesson debrief and revise sessions. As the teach-ers who observed the lesson offer their input, we learn to see the lesson from alternate vantage points. I look forward to building on this start so that we can better serve the needs of all learners.

Choosing Goals Effectively

One of the toughest challenges we faced . . . was coming up with a lesson topic to teach. Our initial goal was to focus on using technology. We soon discovered, however, that the goal of using technology in a lesson made content goals secondary. We were so excited about using technology that we could not decide what topic the technology would be used to teach. We spent much time working like this before we realized that a content goal should be our main focus and the technology should be used in a manner consistent with the content goal. We have decided that in the future we will set a content goal first and then make decisions regarding the technology that would best be utilized to achieve that goal.

Deepening Lesson Study Practice and Learning over Time

It is clear that for a new team, lesson study can be at times exciting and rewarding and at times challenging, and that it can feel like following a set of not-totally-understood steps. But, as a team moves into their second, third, and subsequent cycles, they understand better not just what to do, but why. Team discussions go deeper, the team gains a better understanding of the big ideas of lesson study, and the work of the team becomes more effective. The importance of sustaining the work over time becomes clearer. Teachers also gain a stronger sense of the place of this work in their professional lives. Lesson study begins to feel less like professional development provided *to* teachers and more like teacher-initiated, teacher-led professional learning. It becomes part of our professional role, as educators. And since this all happens over time, the story of deepening the work runs hand in hand with the story of sustaining the work.

One teacher's reflection, after four cycles of lesson study, highlights this sense of professionalism and points to many features of strong lesson study practice that give it its power:

I love Lesson Study. I feel empowered. I was trusted to make decisions about what to teach and exactly how to teach it. . . . I feel like this level of trust in teachers' educational and mathematical expertise is a piece of professionalism that is sorely missing.

I also love that Lesson Study focuses on students' learning. I agree with the author . . . who said that Lesson Study gave him/her new eyes with which to see the students (Lewis 2002a). I definitely feel that Lesson Study has given me more than one pair of new eyes. The lessons that [we] worked on have clear learning goals. To observe students so carefully through the lesson provides insight into what is actually happening in students' minds. I appreciated the opportunity to just be an observer of students, rather than the teacher. When relieved of the overwhelming responsibilities of teacher, I learned more about students each time I observed them. The role of researcher was another highlight of my Lesson Study experience. I found the level of detail in planning, observing, and then analyzing a unit of lessons to be extremely intellectually satisfying. It was dirtying my hands in the real stuff of teaching. It felt meaningful.

Lesson Study also has aspects that just make sense, like collaboration and continuous improvement. . . . Many heads are always better than one. Each teacher's load is lighter when shared with colleagues. Also, the ongoing nature of Lesson Study is an ideal way to incorporate the latest research on education. Every time there's a new educational research publication by a big name, there follows a short-lived reform effort and a multitude of workshops to tell teachers how to use this research in their classrooms. This has never worked effectively. Lesson Study is such a natural way for teachers to use any new research findings, and also to validate or criticize those findings. The idea is that there's always room for improvement—that there's always something more to learn about students and how they understand. It's the antithesis to the quick fix. I want to continually reflect and better myself as a teacher, and Lesson Study provides a systematic way to do just that.

Not all teachers would say, "I love lesson study" as this teacher does, but reading this reflection we can see that over the course of four lesson study cycles this teacher has developed a strong sense of what the work is about and why it is worth continuing. There is little in the reflection about the mechanics of the process and a lot about the larger themes that make the work rewarding and effective. But *how* does this transition happen? How does a team's work grow stronger and more sustainable? And *what* does this more sustainable or robust lesson study practice look like? What changes occur in the team's work or individual teacher's thinking as they gain more experience with the process? How does the team begin to incorporate lesson study into their ongoing professional practice? What is the role of leaders in supporting teams in this deepening of their thinking and lesson study practice? It is the goal of this *Leader's Guide* to help you to answer these questions.

▌ Closing Remarks for Part I

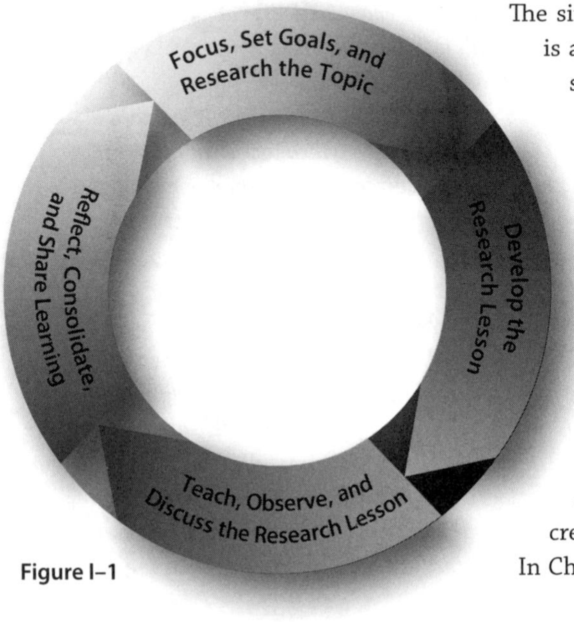

The simplicity of the lesson study cycle diagram is a blessing because it so clearly conveys the steps of the process. (See Figure I–1). Teachers can easily imagine real benefits emerging from this cycle of inquiry, gather a group of colleagues, and start doing it. What the diagram does not convey, however, is that teachers might want to participate in lesson study throughout their careers, gradually developing a more and more powerful professional learning practice. Nor does the picture of a single cycle convey that this practice, given time and support, could operate in a systematic way over multiple cycles to create widespread improvement in a district. In Chapter 1 we have suggested some of the big

Focus, Set Goals, and Research the Topic

Develop the Research Lesson

Teach, Observe, and Discuss the Research Lesson

Reflect, Consolidate, and Share Learning

Figure I–1

ideas and themes that serve as guiding forces in helping experienced teams deepen and broaden their work in these ways, and that can also help novice teams to start out with a strong vision.

In the *Leader's Guide* Part II, we shine a spotlight on each moment in the cycle, attempting to answer the many questions we as coaches have been asked, and to help leaders and teams avoid some of the mistakes we all have made as lesson study novices. By focusing in on each phase, we can see more of the complexity and power of the lesson study process. The themes raised in Part I should come into more clarity as the reader sees how they play out concretely during the cycle. Keeping those themes in mind—research, mathematics, students, and professional community—while delving into the nitty-gritty of what it means to develop and teach a research lesson should provide teams and leaders a good balance of understanding and practical skill. The remaining factor is time—taking time *during each cycle* to think and learn, and taking time *across many cycles* to gradually develop our lesson study practice.

PART II

Deepening Lesson Study Practice Across the Cycle

Reflect, Consolidate, and Share Learning

Reflect, Consolidate, and Share Learning

Develop the Research Lesson

Teach, Observe, and Discuss the Research Lesson

Teach, Observe, and Discuss the Research Lesson

PHASE 1 OVERVIEW | Focus, Set Goals, and Research the Topic

The beginning of each cycle of lesson study is an exciting time. It is the opportunity to establish the tone for your team's work together and to do the initial exploration and research that is necessary to plan a research lesson. Setting norms and goals to guide your team's work and choosing and researching a mathematical topic are the main work of Phase 1 and lay the groundwork for the rest of the cycle. Your team's choices during this part of the cycle are guided by what members of the team already know and by what they find out about their own students' needs. The team chooses questions they want to investigate about lesson content and about ways of teaching that content. They explore and define the mathematical territory of the lesson content and of the unit in which the research lesson sits. In short, the team launches what can be an extended and possibly powerful professional collaboration and learning experience.

This section of the *Leader's Guide* contains chapters that will help you consider team leadership and norms for working in groups, processes for setting goals to guide the team's work, selecting a mathematical topic to investigate through a research lesson, and researching the mathematical trajectory for learning that topic. The set of activities and questions that are central to the team's work in Phase 1 are described in the chart that follows, with notes about chapters for further reading.

Key Activities	Central Questions	Related *Leader's Guide* Readings
Get organized (i.e., schedule meetings, establish norms, seek support, learn more about lesson study)	• *How do we want to work together as a team?* • *How will our research contribute to our school and professional communities?*	Chapter 2, *Team Leadership and Group Norms*
Set broad goals and content understanding goals for students	• *What are our students' greatest needs?* • *What understanding do we seek in the lesson?*	Chapter 3, *Goals and Research Themes*

Key Activities	Central Questions	Related *Leader's Guide* Readings
Select a mathematics topic for the research lesson	• *What topics do students have difficulty learning? Why?* • *What mathematics do we, as teachers, want to explore and learn about together?*	Chapter 4, *Topic Selection*
Share expertise and teaching experiences on the research lesson topic	• *What are our questions and theories about teaching this topic?* • *What is our vision of a good lesson?*	Chapter 5, *Topic Study*
Study mathematics, standards, curriculum, and textbooks to broaden team knowledge of the topic and to outline typical student learning trajectories for the topic	• *What is our students' prior understanding of this topic?* • *How do students' typically learn about this topic?* • *How does this topic connect to other mathematics students will encounter?*	Chapter 5, *Topic Study*

Leadership Focus in Phase 1

Leaders working with teams in Phase 1 should first and foremost assist teams in engaging with the activities and questions listed in the chart, using the chapters in this *Leader's Guide* as a resource. In addition, a few areas of focus for leaders during this phase as a whole include:

- Building trust

- Encouraging teacher ownership of goals and research questions

- Building knowledge of the lesson study process

- Developing shared leadership routines for the team

- Providing access to resources

- Understanding the different types of goals

- Providing expertise in research, content, and study of textbooks

- Encouraging mathematical activity by the team, for example doing problems together

CHAPTER 2 | **Team Leadership and Group Norms**

Lesson study is a process, a cycle of research, a way of learning and problem solving around questions about teaching and learning. Most of this *Leader's Guide* is about what that process is, how teachers begin, deepen, and sustain it. Integral to that process is the *lesson study team*—a collaborative working group. In thinking about starting and deepening lesson study practice, we have to consider the team as well as the process: what kinds of things help the team function and develop strength as a professional learning community? And keeping in mind that *the team is the teachers*, we have to think about how individuals contribute and benefit in this community.

In this chapter, we will treat two areas of team functioning or identity that seem particularly important in building a learning community in the lesson study context.

- *Team Leadership:* What kinds of leadership are needed? How is internal team leadership capacity developed? How does leadership support development of trust, ownership of the team's research, and commitment to team goals? How do coaches or other outside leaders play a role in supporting the team?

- *Norms:* What is the reason for having group norms? What kinds of norms do teams choose? How can they help a team working together productively? Are they really necessary when the team is already used to working together well?

Leadership

The leadership picture for lesson study has an important element of simplicity to it. Lesson study is better described as a community of learners, than as a group of leaders and participants organized hierarchically. There is a culture of learning that extends to all participants. Whether you are a third-grade teacher, a school superintendent, a high school geometry teacher, or a PhD mathematician, when you participate in lesson study work, you come to the table as an equal. You may bring a different expertise to contribute, but you come to learn from the students and from each other.

People from many different leadership backgrounds work with lesson study teams: teacher leaders, coaches, mathematics specialists, department heads, district coordinators, principals, superintendents, university faculty, lesson study experts, and so on. Given the range and number of leader types represented in the list, you might be wondering how *many* of these people need to be involved with a typical team, what each person's role might be, or how teachers fit into this picture.

For most experienced lesson study teams:

- *Teachers on the team are the main leaders of the team's work.* Lesson study is a model for collaborative learning, generally conducted by small groups of teachers who direct their own team activity, and usually share leadership tasks as a community of equals.

- A *few local supporters are closely involved with the team.* Teacher colleagues, a mathematics coach, the department chair, the principal are a few examples of local supporters who help in a variety of ways.

- *The team is connected with a few outside experts who play a more circumscribed, but powerful, role in supporting team learning.*

The particular constellation of individuals and what exactly they do varies greatly from team to team, but the leadership and support they provide falls into five main areas. How leaders provide support in these areas is a main focus throughout the *Leader's Guide*.[1]

- Bringing new knowledge and expertise

- Offering new or different perspectives

- Building professional community

- Supporting implementation and advocating for the team's work

- Facilitating discussion and developing team ownership

Team Leadership

Given the nature of lesson study as a learning community, leadership is likely to emerge directly from team members and be based on a common commitment to the work. Hence, as we noted, most experienced teams largely direct their own work. We offer a few different models for how leadership is shared and evolves, as a lesson study team matures and develops internal leadership and ownership of their work. These models are:

- Teacher-led teams

- Teams led by a mathematics specialist, lesson study coach, or school administrator

- Leadership from outside the school district

---

1   All chapters raise leadership issues and provide ideas for leaders. Chapters that have a primary focus on leadership include: Chapter 2, *Team Leadership and Group Norms,* Chapter 13, *Sharing the Learning,* Chapter 15, *Building Partnerships with School Administrators,* and Chapter 16, *Incorporating Expertise from Outside the Team.* Points in the lesson study cycle where leaders play special roles are detailed in Chapter 3, *Goals and Research Themes,* Chapter 9, *Observing the Research Lesson,* Chapter 10, *Post-lesson Discussions, and* Chapter 17, *Public Lessons: The Lesson Study Open House.*

Teacher-led teams. What does it mean for a lesson study team to direct its own work? For new teams this often means simply adopting an egalitarian philosophy of "everyone shares the work." This philosophy may imply that no one is in charge, but even novice teams realize that there are advantages to having a meeting leader. Rotating this role, along with other jobs, is the usual result. This team meeting leader role is usually about the business of having a productive meeting—seeing that the meeting agenda gets accomplished, that the meeting starts and ends on time, perhaps noting decisions, starting and concluding meetings with statements meant to help focus the group on immediate tasks, and so on. Some groups ask one teacher to be the regular leader—because of his or her special knowledge, skills, or position of trust in the group. This leader may be a teacher who has already participated in lesson study and is mentoring a novice lesson study team.

As the team deepens its work over time, team members will contribute in all areas of leadership, and understand the importance of reaching outside the team for additional support.

- *Facilitation.* The team moves beyond defining the team meeting leader role as keeper of the agenda. Facilitation becomes more substantive. It includes listening carefully and helping the team conduct orderly discussions in which everyone has a voice, keeping the work moving forward, supporting the team norms, and keeping team goals at the forefront.

- *Knowledge of content/pedagogy.* Teachers on the team will contribute resources and expertise, ask guiding questions, and keep track of the bigger mathematical picture—all of which enrich the team's experience and learning. There may be a few teachers on the team who are great at pushing the mathematical conversation, or someone with strong vision and skill in teaching through problem solving.

- *Vision.* The leader should be a person who has a strong sense of the purpose of lesson study and potential of the team's work and can share this vision with the team at critical moments, energizing and focusing the team, advocating for administrative support, helping the team envision and build community.

- *Connections.* As the team gains experience with lesson study, and with leadership, the benefits of connecting to other teams within and outside the school or district become clearer. Members of the team can foster this connection making, widening the circle of expertise shared with the team and the circle of impact of the team's work.

All of these types of valued leadership can come from teachers on the team or from a leader/coach from outside the team. Some leadership roles may be assigned or rotated, but many will be provided in a fluid, informal manner, changing from meeting to meeting as the situation demands and as talents and knowledge allow.

The benefits of the teacher-led team are numerous: the team is independent, develops leadership skills among members, and focuses on issues that are important to them. In other words, the team develops ownership of their work. Most important, these teams have the potential to continue their work even if any outside support is removed. There are also some challenges associated with having teams be completely led from within—especially for novice teams. There may be initial difficulty understanding the lesson study process, and team members may not want to be a leader when they don't know what is even going to happen. Limited contact with outside leaders may limit the teachers' learning of content by restricting the team to its resident expertise. However, having no outside leadership could also force team members to seek fresh viewpoints in other ways—perhaps by inviting colleagues to attend key meetings, or by appointing a team member to be the "outsider" in a discussion. Sometimes there is also a lack of awareness of the team's work in the school (or lack of support from administration) when a team is operating fully on its own.

Teams led by a mathematics specialist, lesson study coach, or school administrator. As mentioned earlier, the cluster of leadership for most teams includes some additional support from a few local leaders, most often a district mathematics coach, mathematics specialist, or department head. For some teams, one of these people plays a major leadership role. A mathematics coach or other mathematics specialist may be the first person in the school to hear about lesson study, the first to initiate and support lesson study teams. The coach may have experience facilitating professional development and working with teachers on classroom instruction and so feel comfortable leading the team. In addition, the coach may see lesson study as supportive of their broad goals for improving math teaching and learning. A coach or department head may participate regularly as a full team member, or may meet with the team periodically, providing resources and leading some of the team's meetings. Providing this kind of close support means taking on a new kind of role as *lesson study coach*. A number of the chapters in Part IV of this book are written by lesson study coaches, or reveal the impact of having support from a coach.[2]

The benefits of this model include potentially strong content and pedagogical expertise and facilitation skill of the leader, as well as stronger alignment of the team's work with school mathematics initiatives. These leaders are also often able to play the role of outsider who brings fresh perspectives to the discussions. Ideally the coach will focus heavily on helping teachers define their own agenda and build internal team

---

2   See, in particular, Chapter 22, *The Longmeadow Story: A District Lesson Study Initiative;* Chapter 23, *Our Lesson Study Journey at King's Highway Elementary;* and Chapter 24, *Expanding Lesson Study Practice at Our High School.*

leadership, and will shift to a less intensive form of support as time goes on. Challenges associated with this model include failure to recognize and develop teacher leadership of the team. This can lead to a situation in which teachers feel that it is not totally safe to voice opinions that go against the leader's ideas, or simply aren't able to continue if the coach isn't present. Another challenge for these leaders is to recognize that lesson study leadership depends heavily on the knowledge, skills, and sensibilities they bring from their prior leadership work, yet requires some new goals, skills, approaches, and assumptions.[3]

Leadership from outside the school or district. As teams gain understanding of the lesson study model, they realize that bringing in fresh perspectives and expertise from outside the team is highly beneficial and should be a regular part of the process. Often outside educators participate at the teaching and post-lesson discussion, or during the team's topic research, but sometimes an outside organization provides major and ongoing leadership support. Some schools conduct their lesson study work through a grant, university partnership, or research study that provides regular team facilitation and training. A district might also bring in lesson study experts to launch lesson study in the school—and these consultants may provide ongoing or periodic leadership or individual teams, or provide workshops for multiple teams.

Outside advisors can offer teams in-depth knowledge of lesson study, content and pedagogical knowledge, connections to other lesson study teams, and access to resources that the team might not have. Challenges include the time limits associated with the funding source, and the potential for not developing internal leadership. Much more will be said about incorporating this kind of external leadership and expertise in Chapter 16, *Incorporating Expertise from Outside the Team*.

Development of Leadership Capacity

When thinking about leadership in a lesson study team, it's most important to consider how the leadership contributes to the development of a strong professional learning community. Initially, it may be helpful to have the assistance of an outside leader or facilitator to foster community and leadership within the team, enrich the content knowledge of the team, or to help the team understand lesson study. Over time, the leadership capacity of the team as a whole can develop to a point where the team can operate independently—bringing in outside expertise at critical moments in the process. It is worth noting, also, that *individual* teachers and coaches on the team are developing leadership capacity through their lesson study work. The process gives participants a chance to develop new knowledge and skills, an increased sense of efficacy, as well as a new sense of teachers' leadership role in the reform of classroom instruction.

---

3   See Chapter 15, *Building Partnerships with School Administrators* for more on this challenge, and more detail on the role of school leaders in supporting teams.

▌ Group Norms

Group norms are another critical factor in team functioning and in the development of professional community. Teams that are composed of old friends might feel they don't need them. Teams that don't know each other well may find it a lot easier to start off by jumping right into the work. However, since lesson study offers such a different model for teacher collaboration than most professional development, department meetings, or committee work, teachers can expect to be working together in new ways. Teams can build on the positive foundation of their existing norms—ways of working together that have worked well in the past—but will also need to think about new norms that will serve them well in this new context.

We encourage all teams to talk about their expectations about lesson study work, purposefully choose norms, and plan to check in on them periodically. Norms can help the team create a professional and safe culture for working and learning in this new context, overcome common challenges, come to consensus, and ensure that work gets done efficiently. Perhaps because the day-to-day work of teachers is largely individual and collaboration may occur primarily on short-term tasks (like creating a common assessment), setting and monitoring norms may be a new way of working. It may take the encouragement of a coach or teacher leader to get the norms conversation started and to come to consensus around a few.

> A new team might begin setting norms by sharing ideas on the question, "In your experience, what makes a group work well or poorly?" This easily starts the conversation and may suggest a few basic norms for the team. Existing teams may ask, "What changes in our norms might help us improve our work?"

A typical set of norms chosen by a new team includes things about *hearing everyone's viewpoints*, *starting and ending meetings on time*, *keeping notes about decisions*, and *sharing the work fairly*. As a team moves through multiple cycles, toward a robust and sustainable lesson study practice, team members will take on greater ownership of the norms, and the norms will be more specific to lesson study. Teams will begin to use the norms not only to foster smoother operation, but to solidify lesson study practices that support team learning, and to overcome challenges. Leaders can play an important role in helping teams make these shifts.

In thinking about how your team can move toward this deeper level of norms, consider the following areas:

- Routines and protocols that guide the flow of the team's meetings and cycle

- Practices that promote team learning

- Structures to support the team's decision making

Each of these areas is described briefly here to help you and your team think about how you want to establish a professional community that functions well, builds trust, and supports learning about the team's goals.

Routines and Protocols That Guide the Flow of the Team's Meetings and Cycle

Lesson study teams have a lot of work to do. Establishing routines is one way to help the team function smoothly and get the work done. Routines usually include the ways a team shares the work, keeps records, and uses agendas. Many teams also have routines for celebrating accomplishments and providing refreshments. Components of the lesson study process that are very new for teachers are also well served by routines and protocols. This includes the observations and post-lesson discussions, as well as the team's integration of mathematical explorations into their work together.

Protocols for Observations and Post-lesson Discussions

Typical observation protocols[4] define the expectations for observers at the teaching of the lesson and for participants in the post-lesson discussion. These protocols help newcomers to lesson study understand how to observe in this new way and summarize the type of professional conversation that is expected. More detail about these protocols is given in upcoming chapters.[5] However, there is no one correct protocol. A team will craft protocols over time that support the learning environment of the team and help newcomers understand the culture.

Mathematical Routines

Many teams have developed a routine of "doing math" in their meetings because they learn so much from working on mathematics with colleagues, and enjoy the intellectual challenge so much. One team had three rotating jobs: leader of the meeting, provider of snacks, and provider of an interesting math problem on the lesson topic. The group started every meeting eating the snacks and doing math together. Over time, a friendly competition evolved around providing particularly interesting problems. Another team developed a norm that they would not finalize their choice of lesson topic until the team had completed several challenging extension problems on the topic. More than once, a promising topic was abandoned after these explorations revealed the limits of the mathematics involved. For the team, this routine promoted powerful team learning and improved topic selection.

Practices That Promote Team Learning

To function well as a learning community, teams will want to establish practices that allow for mathematical learning and that foster dialogue about teaching and learning of mathematics.

---

4   Protocols utilized by Japanese teachers for the observations and post-lesson discussions have been used as guides for U.S. teachers who are new to lesson study. See Lewis (2002a).

5   See Chapter 9, *Observing the Research Lesson* and Chapter 10, *Post-lesson Discussions* .

Fostering Mathematical Learning

In order to establish a safe environment for mathematical learning, the team will want to think about the norms that relate to their joint mathematical work. Carroll and Mumme argue that

> . . . leaders need to develop their awareness of the sociomathematical norms that are negotiated in their groups and intentionally work to foster a *robust* set of sociomathematical norms—norms that promote deep understanding of mathematics. These norms involve what and how teachers' ideas are shared, what counts as a mathematical argument, how confusion and mistakes are handled, the nature of questions, etc. (Carroll 2007, 76–78)

If you have sat in a mathematics workshop and felt unsure of how to solve a problem, or embarrassed that your solution was too simplistic or wrong, or if you have seen a colleague with similar doubts or hesitations, you will appreciate the need for these norms. Working on mathematics problems together is a new experience for many teams, and at the same time a central strategy for team learning. This makes a discussion of sociomathematical norms of great importance in lesson study. A bonus of taking time to reflect on and how these norms operate within the team is that it puts us in touch with the same need for norms in our classrooms, where increasingly we hope to foster mathematical dialogue and group problem solving.

Fostering Dialogue

"Two different ways of talking are essential for improving schools. Discussion, in which the goal is to make decisions, is one way. The other is dialogue, in which the goal is to develop understanding. . . . [In dialogue] ideas can be presented, and explored without judgment. Group members seek to understand each other's viewpoints. When dialogue precedes decision-making, decisions are more likely to stay made" (Garmston and Wellman 1999). The distinction drawn here, and the importance of both kinds of talk, is clearly highly relevant to the specific case of lesson study. Further, these authors have suggested "seven norms for collaborative work"[6] that are specifically intended to foster dialogue. These norms include, for example, pursuing a balance between advocacy and inquiry, and presuming positive intentions. Reading about and discussing these would make an excellent starting point for an experienced team that wants to review and enrich how the team talks together during lesson study.

Structures to Support the Team's Decision Making

Figuring out effective ways to settle the multitude of small and large decisions that go into choosing a focus and developing the research lesson is probably the most universal

---

6   The norms include: Pausing, Paraphrasing, Probing, Putting forward ideas, Paying attention to self and others, Pursuing a balance between advocacy and inquiry, and Presuming positive presuppositions.

challenge that teams face. What kinds of situations might lead a team to add decision-making norms?

Unequal Power or Expertise

It is not unusual for the least experienced, least knowledgeable, or least vocal members of a team to withhold their input at critical decision points, due to lack of confidence, or a multitude of other reasons. Similarly, vocal team member(s) may assume things have been decided before everyone has weighed in. In the long run, this phenomenon could lessen individual teachers' sense of ownership over the work and lower the quality of the team's research because fewer ideas or viewpoints have been put on the table.

Decision Making That Is Impeding Momentum

Sometimes a team is great at brainstorming multiple options for their lesson but much less able to settle on one. The level of creativity and energy about the ideas is probably extremely rewarding to the team, but such a team may realize after a few cycles that they need a way to determine when it is time to make a decision and move on. Solutions might include creating a timetable for key decisions, or norms to support steady progress. Teams facing this dilemma might also do well to remember that all observation lessons are, in effect, experiments. The goal is to agree on ideas that seem to be worthy of experimentation, not to decide everything conclusively through discussion.

Major Stalemates

If a team simply *cannot* agree through consensus on some key question, it can be uncomfortable in the least, and frustrating or destructive at worst. Encountering this kind of impasse even once is likely to inspire the team to find ways to avoid or break such stalemates in future work. This may reflect a breakdown in the team's general norms, and suggest that revisiting norms is a good first step. If it is purely a tiebreaker that is needed, taking a team vote or gathering some student data to inform a choice might work. Revisiting the mathematical understanding goals for the lesson is often the best way to resolve such stalemates. In some teams, the teacher of the lesson has final say in stalemate situations. This strategy needs to be used carefully, however, so that it does not sabotage any essential ideas that the team as a whole has established for the lesson. Dr. Akihiko Takahashi, an internationally known lesson study researcher and practitioner, suggests that teams remember the following saying, "When the team cannot decide, ask the students." In other words, teachers could debate theoretically for a long time and remain unsure. What we want to know is how students will respond to the lesson, which we can only learn by seeing the students in the lesson.

"Do We Have Norms?"

Ironically, it is not unusual for teams to set norms, and then subsequently ignore them. Finding a way to truly activate their norms is another common challenge for teams.

Reflecting together on why norms are being ignored could be quite illuminating. This discussion might produce a decision to have a "norm to revisit norms"—perhaps one like the following.

- Include norms check-ins on meeting agendas or as part of leader's role.

- Check in on one norm at the end of each meeting.

- At the beginning of every cycle, make sure we have useful norms.

- Include norms among our topics for end-of-meeting team reflection and discussion.

Failing to use or maintain the group's norms is a common challenge because these norms often represent a cultural shift for the team, or are very new practices. Earlier, we described norms as more like expected ways of working together than as rules for how the team will operate. Yet when norms are newly chosen, and represent very new ways of operating, they may need to temporarily take on a more prescriptive status. The longer the team works at them, the more the team can refine them and incorporate them into their standard ways of working together. Even in this situation, though, revisiting the group's norms is important to ensure that they represent the changing goals and membership of the team.

Leadership and Norms in Professional Community

At the beginning of every lesson study cycle, but most clearly in a first cycle, a team has the opportunity to define the work it wants to do and the type of learning community it wishes to build. In other words, the team establishes the content of its work and its goals for the upcoming cycle and for the long term. By making decisions about the team's leadership and norms for work and discussion, the team is saying, "This is the type of community we wish to be." This is a powerful first step in establishing ownership over their work, and building commitment to the team as a learning community. The next step in this same trajectory is to choose a research focus and connect it directly to the issues the team faces in teaching, and to students' greatest needs. This goal setting, which is the subject of the next chapter, is a further step in establishing ownership and commitment to the work.

CHAPTER 3 | **Goals and Research Themes**

Goal setting is a critical activity in lesson study because the team's goals form the foundation for their work throughout the cycle. The first step of lesson study is to ask, "What is a problem we wish to solve or a question we wish to explore about mathematics teaching and learning?" and to frame a set of goals accordingly. Teams adopt three kinds of goals:

- Broad goals for students' development as learners

- Unit- and lesson-level goals for student understanding of mathematics

- Goals for teacher learning

These goals bring focus to the team's work throughout the cycle. "Will this choice help achieve our goals?" becomes a mantra. The goals also define the mathematical content area the team will be exploring. Goals also matter in terms of the relevance of the work to the team. In lesson study, teachers have ownership over the process of creating and working toward these kinds of goals. Teachers' excitement with the lesson study process is often due in part to the opportunity for systematic investigation of goals that *they* have chosen, and that matter for *their students*. Setting and working toward goals is a tremendous bonding force for teams and provides strong motivation for their collaborative work.

In this chapter we elaborate, with examples, on each type of goal, stressing how teams choose them, what they offer teams in terms of learning opportunities, and how leaders and teams can think about deepening their work by using goals effectively.

Broad Goals for Students

Our original goal-setting involved an extraordinary number of hours . . . but what we were really doing is establishing a shared expectation of what makes a strong student and strong teaching. I think, for us, it was time well spent. A good overarching goal should be focused on the student—what kind of person/student do we want to develop? Our overarching goal—to develop curious students, willing to take risks, who aim for skill mastery . . . now pervades our thinking when we talk about teaching. A common question now is "Is there a way we can tweak this lesson so that it is the student doing the thinking?"[1]

---

1   See Chapter 18, *The Importance of Goals in Lesson Study* to learn more about how this team set their goals and what the long-term impact was on their lesson study work.

Typically, teams first choose one or two broad, overarching goals for students. These broad goals represent the qualities that the team wants to develop in their students as mathematical learners. Broad goals frequently address persistent challenges to students' learning, are grounded in students' greatest needs, or tackle especially important skills or knowledge. Teams try to avoid choosing too many of these goals because goals are meant to focus the team's thinking and research. Examples of broad goals for students are:

- We want our students to use multiple strategies for solving mathematics problems.

- We want to develop students who have confidence in their work.

- We want our students to be independent thinkers—able to derive individual approaches to problem solving.

- We want our students to see that working with classmates can deepen their understanding of mathematics.

In Japan, where lesson study originated, teachers frequently set broad student goals that they then work on for multiple years. In addition, broad goals are often set and shared by the faculty of a whole school and serve as a school research theme. Fernandez and Yoshida describe this as follows, "it is typical for a school to maintain the same *konaikenshu*[2] goal for a period of several years (Lewis and Tsuchida 1997). This prolonged focus is meant to provide enough time for the school to make significant progress in moving closer to attaining its chosen goal (see also Maki 1982). It is not unusual, however, for a school to focus on different aspects of its chosen goal, or to take different perspectives on this goal, from one year to the next"[3] (Fernandez and Yoshida 2004, 13). These broad goals serve as vehicles for whole-school improvement and change. At the end of the year, schools will often produce a written report that summarizes their lesson study work, delineates their learning, and attempts to measure progress toward their broad goal.

Here in the United States, whole-school participation in lesson study is not yet common, but a team's broad goals can still serve as research themes that bring a unifying focus to their research over several cycles. Working on a research theme over time is more possible for experienced teams as they deepen their lesson study practice, and is

---

2   A Japanese term for in-school professional development.

3   Fernandez and Yoshida 2004. *Lesson Study: A Japanese Approach to Improving Mathematics Teaching and Learning.* Mahwah, NJ: Lawrence Erlbaum Associates, Inc. p. 13. Referencing Lewis and Tsuchida 1997. "Planned Educational Change in Japan: The Shift to Student-centered Elementary Science." *Journal of Education Policy* 12(5): 313–331, and Maki M. (Ed.). 1982. *Kyoin Kenshu no Sogoteki Kenkyu [A Comprehensive Study of Teacher Training].* Tokyo: Gyosei.

certainly not an expectation for new teams. As lesson study becomes more established in a school, teachers may identify schoolwide, department-level, or cross-department goals. In these cases, individual lesson study teams work on the same broad goals but apply them to the context of their own subject area. For example, if the school's broad goal is to develop students' communication skills, a team of mathematics teachers might choose convincing mathematical arguments as a specific area of mathematical communication on which to focus.

How Do Teams Choose Their Broad Goals?

In the United States, teams often begin the process of developing broad goals for students by taking time to talk together about who their students are and what their greatest learning needs are. In these discussions, teachers learn that they have many of the same hopes for their students, and come together around a few common challenges they would like to work on. A brainstorming activity that many new teams use to start this discussion begins with the questions: Who are our students? What are their qualities as learners and as people?[4] Teachers brainstorm two lists: (1) the type of mathematical learner they hope their *students will become*, and (2) what *students' current qualities* are. Here are some examples:

We would like our students to:

- Take responsibility for their own learning and persevere.

- Be proud to be good at math.

- Make connections among math ideas.

- Be willing to share thinking even if it might turn out to be wrong.

Our students now:

- Are social and care about their peers.

- Think math is all about procedures.

- Are creative and like things that require puzzling.

- Give up easily on challenging problems—prefer to be shown how, or given the answer.

Through discussion, teachers identify those qualities that have the greatest importance for their students, or that represent the greatest disparity between the current and ideal and incorporate them into a team goal. While the brainstorming activity always

---

4   This goal setting activity is described in Lewis (2002a, 56–57).

produces a long list of potential goals, it is important for the team to reach consensus and choose just one or two on which to focus.

Most teams will stay with their broad goal(s) for more than one cycle, because the problems they are focusing on are complex and addressing them in a systematic way over time makes the most sense. When teams stay with their goals for several years, they revisit them periodically. Post-lesson discussions and end-of-cycle team reflections will highlight what the team is learning about their goals, and help the team to assess their progress on them. At the beginning of each new cycle the team can discuss whether the goals are still relevant to student needs and compelling to the team.

Lesson Goals for Student Understanding of Mathematics

> *Identifying the learning goal and staying focused on that goal [was one of the most rewarding parts of the process]. It was easy to go off on tangents but having the goal up front and in writing kept bringing the lesson back on track. . . . A big moment was when [our coach] stopped us from jumping to [planning] activities before we completely investigated our goal and its implications!*

In writing daily lessons or in preservice training, many teachers have been required to state lesson goals as behavioral objectives, for example: "The student will be able to calculate the area of a triangle." For a *research lesson*, this kind of goal may not be the most productive, because teachers aim to learn more than whether students can successfully perform mathematical tasks. They want to learn about *how* students understand and learn mathematical ideas. What are our students thinking? What do and don't they understand? What tasks will help them develop greater understanding? Therefore, lesson goals for a research lesson should articulate not just what a student will be able to do, but what understandings she will have. For example, a team focusing on slope might state a goal like, "Students will understand slope as a rate of change." Pushing their thinking further, the team should make sure they are clear what this understanding actually involves, by discussing, "What does it mean to understand slope as a rate of change?" This may be a very hard question to answer, but trying to achieve clarity on it will make a great difference in the quality of the research lesson and in how much the team is able to learn from it.

It can be very challenging to come up with understanding goals that are neither too broad nor too narrow. For example:

- Too broad: *Students will understand linear equations.* From this goal, it is not clear *what* we want students to understand about linear equations, and the goal may suggest a very large unit of study.

- Too specific or procedural: *Students will identify the slope and y-intercept of a line given an equation in y = mx + b form.* This goal might assess student knowledge of terminology, but tell us little about their understanding of lines and slope.

- A better goal: *Students will understand the connections between the graph of a line and its equation.* This goal is about understanding multiple representations rather than focusing solely on procedural skill, and identifies a specific area of understanding that could be the focus of a lesson or small unit of instruction.

How Do Teams Develop Their Student Understanding Goals?

The process of determining the lesson understanding goals begins during topic selection and research.[5] Team members begins by pondering and discussing their own understanding of the topic. Continuing with our example on the topic of lines and slope, teachers might ask themselves: "What does the word *slope* mean to me? What do I know about it?" The discussions may surface ways of thinking about the topic that teachers have held for so long that they don't even think about them anymore, for example, that slope is related to incline. These fundamental assumptions may not be held by students and may need to be addressed in the lesson goals and design.

Textbooks are another support that teams can use to help think about understanding goals. Although some texts state only procedural goals, many of the new standards-based curricula have clearly stated lesson goals that emphasize understanding, and mathematical notes for the teacher that pinpoint the kind of understanding the lesson is aiming for. Secondary texts often include proofs of theorems, or explanation of key concepts, that the team can study to clarify *what there is to understand* about the topic.

The understanding goal is honed further during lesson development as the team sketches out the flow of learning across the unit, and comes face to face with what exactly students might be able to accomplish in a single lesson, or exactly what kind of thinking the lesson activities will promote. And a last refinement may occur as a result of the first lesson observation and post-lesson discussions of evidence.

The work of crafting good lesson goals, and seeing this work as an opportunity for the team to build their own understanding of the topic, is an area where teams can deepen their work over time. A good first step in that direction is simply to look at the wording of various lesson goals, and think about exactly what understandings each one implies.

A Note on Assessing Student Learning

In lesson study, there is no shortage of opportunity to assess student progress toward the content learning and broad goals. Many teams include written assessments or

---

5   More details on this process are found in Chapter 4, *Topic Selection* and Chapter 5, *Topic Study*.

student reflections on learning in their unit plans. The lesson is usually designed to reveal student thinking and misconceptions, making the research lesson observation and post-lesson discussion prime opportunities for assessing student learning. Multiple observers will be looking specifically for data on student learning about all of the team's learning goals, and will share and make sense of the evidence during the post-lesson discussion. Artifacts from the lesson (student work, end-of-lesson student reflections, video, and so on) are also analyzed, and as the cycle ends, the team takes some additional time to think together about what they have learned about their goals.

Goals for Teacher Learning

Goals for student understanding and learning—both broad and specific—are certainly important, but so are the team's goals for what they will learn as teachers. Participants in lesson study view themselves as lifelong learners, and the lesson study cycle enacts this philosophy as it combines the experimental process of developing, observing, and analyzing a research lesson with opportunities for studying mathematics and pedagogy related to what teachers know are difficulties for students. Goals for teacher learning are too easily left implicit, so an important role for leaders in the lesson study process is to encourage the team to articulate and define what members hope to learn, and to return to those goals throughout the cycle to determine what the team has indeed learned.

Questions such as the following at the beginning of the cycle can help a team think about *what they would like to learn*, and therefore can help determine the trajectory the team will follow as it sets goals for students, chooses mathematical content for the research lesson, and designs the lesson:

- What area in mathematics do we wish we knew more about?

- What topic in mathematics have we always struggled to help students understand?

- Which pedagogical models or strategies would we like to understand or apply better in our classrooms?

- What focus areas do schoolwide student challenges or assessments suggest?

- What research on mathematics teaching have we read that we would like to explore with our students?

The primary goals for teacher learning are expressed in the *research questions* the team is exploring through their lesson. For example, a team developing a lesson on fractions might want to learn:

- How do students think about fraction multiplication?

- What concrete experiences help them understand this idea?

- How does student understanding of fraction multiplication build on (or challenge) their prior understandings of multiplication?

Many lesson study teams choose additional learning goals that are team centered. Here are a few examples:

- A team may state, "Our goal is to learn about calculus, because our knowledge of it is not sufficient to teach our precalculus classes well." This team may then choose their lesson topic to suit their team-learning goal, and seek outside expertise to assist them in building their own understanding before and during the cycle.

- Many teams choose a team-learning goal around improving their understanding of lesson study. For example, "Our team-learning goal is to improve our observation and post-lesson discussion process." During the cycle, the team may plan to read an article, and attend a public lesson at another school.

- Teacher- or team-learning goals may extend across many cycles and relate to the makeup of the team. For example: "Our goal is to learn about the development of algebraic ideas across the grade levels taught by our team (grades 5–8)." This goal might push the team to spend time researching state or national standards, or reading research on teaching algebra in middle school, or developing lessons that are presented in more than one grade.

- Other team-learning goals are related to the team's participation in a schoolwide initiative, or designed to spark a schoolwide initiative. For example, one team's learning goal was to find ways to integrate mathematics across academic and vocational classes in their school. The team's lesson study work was the initial step in opening up schoolwide dialogue and changes in teaching. (See Chapter 21, *The Essence of a Day: An Open House Story* to learn more about how the team's goals impacted the teachers and the school.)

Continuing to ask questions about teacher learning throughout the cycle can help focus the team on *what they have learned* (in contrast to what their students have learned). For example, the question, "How have my ideas changed about how students learn this topic?" would be appropriate after the team has studied various textbook treatments of the topic, after they have observed students during the research lesson, and again during end of cycle reflections.

A Note on Assessing Teacher Learning

Assessing progress on the teacher-learning goals is also built into the lesson study process. This assessment is usually not a test, but a series of informal reflections and summative reporting on the research. During the team's topic research and lesson development, teachers have many opportunities to talk about what they are learning. For example, many teams take a few minutes at the end of each meeting to reflect on their learning. The post-lesson discussion, though not designed to assess teacher knowledge, does provide the team with important feedback on their learning. Comments from guest observers with strong expertise or fresh viewpoints will inform the team about the mathematical correctness of their lesson, and the accuracy of their assumptions about the mathematical learning trajectories and connections. At cycle end, team members takes time to step back and assess their own learning in a more summative way, and to state goals for future learning.[6]

Lessons Learned About Deepening the Team's Work Around Goals

Whether the goals are broad goals for students, understanding goals for the lesson, or teacher-learning goals, taking time to consider your students' needs (as well as your team's) and choosing goals that reflect them can contribute richly to your team's work over time. Reflecting with your team on the following list of lessons learned, or reading the teacher reflections in Chapter 18, *The Importance of Goals in Lesson Study*, may provoke thoughtful revisiting of your team's goals.

- Our teams have found that the best goals are *grounded in* their *students' greatest needs* and feel *real and useful*, whether those goals emerged from teachers' observations, the school's mission, serious teaching challenges, or analysis of students' achievement data.

- Content understanding goals are not easy to develop, even for an experienced team. It helps to keep asking "What do we understand, or not understand, about the mathematics?" and "What prior understandings do our students have?"

- Choosing goals that are *consistent with an ongoing or new district initiative* often strengthens a team's support, visibility, and connects the work to local issues.

- As team members gain experience, they may see the value of staying with one broad research goal or theme across multiple lesson study cycles, or may find it easier to identify a goal worthy of that long-term investment. Multiple research lessons on one goal give a team (and students) more time to progress on the goal, and during each cycle, the team can learn something about one aspect of the goal.

- Lastly, it is always a good idea for teams to *think about how they will evaluate their goals*. How will the team measure whether it has reached or achieved its goal? What will the team see in students' work if the goal is achieved?

---

6   See Chapter 12, *What Have We Learned?* and Chapter 13, *Sharing the Learning* for more information about this end-of-cycle summary of learning.

Lesson study is fundamentally a goal-driven process. Whenever possible, teams should step back to consider whether the goals are playing a strong role in helping the team make decisions about the research lesson design, in choosing what observation questions to pursue, and so on. Some key points in the lesson study cycle for checking in on the team's goals and whether they are still driving the lesson study process are when developing the lesson, when preparing for the observation, at the post-lesson discussion, and when consolidating and sharing the team's learning from the research lesson. In moments of team conflict or indecision, when discussion seems to have reached an impasse, there are few questions more helpful in resolving conflict or restoring forward motion than asking "Which choice here will best meet our goals?" Over time, teams get better at setting goals and at keeping them at the center of their work. The voice directing attention back to the goals might initially be that of a coach for a novice team, but will certainly be that of a teacher on the veteran team.

CHAPTER 4 | **Topic Selection**

At the onset of every lesson study cycle, your team will choose a content focus—a concept or topic in mathematics that students will be learning about in the research lesson. Selecting the topic is an important opportunity for your team to define what student learning you will impact, what your team will be studying, and what your colleagues and the wider profession stand to learn from your work. Taking some time to research and debate possibilities generally enables a team to make a choice that will allow them to pursue their main goals. The time includes valuable debates about educational goals and philosophy, what the students at the school need most, and pedagogical issues. Discussions like these are some of the most valuable part of the lesson study process. Sometimes the selection of a mathematics topic can seem a bit daunting. One teacher reflected after one of his first lesson study cycles:

> *From my experience, developing a topic seems to be the most difficult part of lesson study. . . . We spent six hours talking about various topics and how we could develop lessons that would elicit student inquiry. Finally we arrived at our topic, and then the quest was to develop the content to achieve the goal of student learning, not memorization and demonstrations. This has been hard . . .*

Clearly, this is much more time than a team typically spends to hone in on a topic, and the team did plan to shorten their process in their next cycle, but their cycle was enriched by the thorough sharing of ideas about mathematical topics that had taken place at the beginning. At another extreme is the team that quickly reaches consensus about a topic to pursue. One coach working with a lesson study team reflected:

> *It was clear almost immediately that the team shared a common thought, "Let's focus this cycle on probability." Their textbooks had no probability units. Their students had done very poorly last year on the "area-model" probability items on the statewide test. And many on the team said that their own knowledge of probability was weak. So the decision was clear—to create a small unit on area models for probability to supplement the texts, and to try to learn about how children at this grade level think about probability.*

How could one team take hours considering topics and another just a few minutes? The issue is not one of finding a topic, because there are many possibilities. Instead, it's an issue of narrowing the field of choices, purposeful selection, and of coming to consensus. Teams must have a process for nominating possible topics, explicit criteria for what makes a good topic, and an effective team decision-making process that allows the team to make the final choice smoothly.

A Nominating Process: Getting Ideas on the Table for Consideration

A simple and effective process for creating a list of potential topics begins with each teacher on the team reflecting *individually* about student needs and the team's learning goals. Following this individual reflection, the team members share their ideas, post all the topic suggestions where everyone can see, proceed to discuss their relative merits, and choose a topic. Using individual reflection prior to team discussion is more likely to elicit ideas from all team members, and to put a range of possibilities on the table. Teams usually find it helpful to use *one or two* questions like the following to spark this initial brainstorm:

- What topics do students have difficulty learning or have deep misconceptions about?

- What topics do teachers find difficult to teach?

- What are the most important mathematical concepts I want my students to learn this year?

- For which mathematics topics do we want to learn more about how students think?

- What mathematics do *we* want to better understand?

- Are there new additions to our curriculum that we want to explore?

- What does our student assessment data tell us about mathematical content where achievement is consistently low?

- Which topics will students be studying at about the point on the calendar when the team hopes to teach the research lesson?[1]

Novice teams will often be most concerned with the last question. The question of when topics are scheduled to be taught during the year is certainly a valid concern, but as teams become more experienced with the whole lesson study process they are often able to become more flexible about how to schedule creatively so they can develop research lessons on topics selected based on other criteria as well.

Criteria for Selection: What Makes a Good Topic?

A general way to think about selecting a research lesson topic is to assume that *a good topic is one that matters*—for your students, mathematically, and to the team. Your team will devote considerable thought and time to exploring this topic and developing

---

1   For some teams, the schedule of team meetings or restrictions in coverage for observing the lesson suggests an optimal date for the research lesson. These teams may look for topics that students will be studying around that date.

the lesson, so you will want to find the topic interesting, see the work on this topic as helpful to your students and your teaching, and be eager to share what you learn with colleagues.

Topics That Matter for Our Students

The initial team brainstorming usually produces a list of topics that matter in some way to your students. But the team will still want to discuss how a strong understanding of each brainstormed topic would benefit students, what makes each topic difficult, and what special student needs in your school point to certain topics. For example, a school focus on making algebra accessible to *all* students might point to topics in algebra that students with limited prior knowledge find most confusing.

Topics That Matter Mathematically

In this context *important mathematically* means that the topic is a foundation for future learning and it helps students make important mathematical connections. The advantage of choosing important topics is that the work of the team can have a wider impact. Most teachers do teach these topics, and so will have an interest in learning about the team's findings. Understanding of these topics benefits students as well because the topics are usually critical elements in the curriculum.

Research lessons on topics of importance can have wide ripple effects. One middle school team explains why they chose place value as a focus topic:

> *A few minutes into our first meeting this year, one of the team members said, "It seems like a lot of student difficulties stem from place value." With that short comment we were off and running! We thought that students had only a superficial understanding of place value and, in particular, the relationship between the different places. Our theory became that new middle school math concepts were made more challenging to students because their view of numbers and place value was not adequately flexible. The number of middle school topics affected by place value is extensive: mental and therefore more efficient math calculations, scientific notation, decimal operations, metric system and determining reasonability of answers all rely on place value.*

For this experienced team, the merits of focusing on place value were clear. They felt that students arrived in fifth grade with a weak understanding of place value, and that this generated difficulties for them in their middle and high school studies—in particular because place value was not included in their middle school curriculum. The team felt that students would be reluctant to repeat approaches they had experienced in elementary school, so they wanted to come up with a new approach for teaching about place value. Their ultimate approach, a small unit with activities in base 5 (i.e., something new for students but dealing with the same place value understandings) was so effective that they decided to use the unit with all their fifth graders at the start of the following year.

Some important topics are *large*, like place value, and extend across many years of school and involve a great range of understandings. Others may be small, but are important as links, connectors, foundations, or powerful ideas. If your team isn't sure which of the topics they are considering are most important, some strategies include:

- Considering the question: *"If my students could only learn five things this year, what would they be?"*

- Inviting a knowledgeable person outside the team to discuss topics with the team.

- Checking the National Council of Teachers of Mathematics (NCTM) *Principles and Standards for School Mathematics* and *Curriculum Focal Points* documents.[2]

- Doing mathematics problems on the possible topics. Working on problems from the target grade level (and from prior or later grades) will not only stretch the team's thinking and understanding of what the topic is, and what students learn at various grade levels, but help them think about how the topic fits into a larger mathematical strand or context.

Topics That Matter to the Team

This work needs to matter to your team, and so team goals and research questions will also strongly affect your topic choice. A vocational high school team explains how they made a topic choice that would help them investigate their research theme—integration of academic and vocational learning:

> As a group we are interested in observing the connections between what students learn in the academic classroom and the need to carry over what they know in the novel settings of agricultural classes. . . . Our group identified the topic of rates as an area where students had a difficult time using their math skills in an applied setting. . . . We first observed a grade 10 math class lesson on the filling of bottles with varied shapes and plotting volume vs. height to begin to develop the connection between slope, changing slope, and the resulting graph. Then we observed an environmental science class where students visited a stream on campus and calculated the flow rate (volume/time) by measuring the depth, width, and length of the stream under the bridge and counting the time it took for a floating object to clear the bridge.

Sometimes the team chooses the topic specifically to further their mathematical learning. For example, several Lesson Study Communities in Secondary Mathematics teams were able to have a university mathematician join their team during one cycle— through a district or research grant. To benefit from this expertise and deepen their own understanding each team chose a topic for that cycle that they wanted to study in depth.

---

2   The Resources Appendix includes some suggestions for resources for exploring mathematics topics.

Other topics that may rise in importance for the team include content areas that are new or have shifted grades in the curriculum. For example if a district moves the study of fractions from fifth to fourth grade, the fourth-grade lesson study team might choose to focus several research lessons on these fractions ideas that could be understood at this earlier stage in students' mathematical development.

Not surprisingly, practical considerations and logistics do often play a role in what topics matter to the team at the particular point in time. The most common of these is timing. Teachers usually find it most illuminating to teach a lesson when it fits naturally in their curriculum, and so will choose topics that are occurring during the dates they are able to conduct the cycle. In this case, teams should record the other possible topics for consideration in future lesson study cycles.

Effective Decision-Making Processes for Final Topic Selection

It is very likely that a consensus on one topic will emerge naturally from discussion. If not, the team will probably narrow the possibilities down to two or three topics, and may find it useful to first revisit their goals and norms:

- Is one mathematics topic better aligned with our goals?

- Which mathematics topic will help us learn the most?

- Is there something about our norms or group process that is keeping us from a decision?

If the check on goals and norms doesn't produce a decision, the team probably needs to step back and assess the situation. What is going on? Are we faced with too many good choices, or are we just not excited by *any* of them? Is our debate interesting and helpful to us—does it give us a chance to talk about philosophy of teaching or importance of content?—or is it just frustrating, tense, or going in circles? If the former, the team can simply take a vote on closure—deciding how much additional debate time is wanted, then moving to a decision. If it is the latter—a negative situation or true stalemate—a new strategy is needed. Two ways for a team to achieve a breakthrough during a stalled discussion or decision-making process are to infuse the discussion with some new source of data or to revisit group processes. The team might bring a new perspective to the discussion by:

- Going back to the mathematics—shift from abstract discussion of the merits of topics to actively investigating the mathematics further, and base the decision on the mathematics.

- Seeking advice from a knowledgeable resource outside the team or reading research to suggest the relative merit of the topics.

- Trying a new consensus-building technique. One that we have seen work quite well is the "silent conversation" during which all team members post brief notes about each topic on poster paper or a blackboard, responding to each others' comments in writing, and in effect having a silent conversation. New ideas often come up that clarify or change people's opinions, but also, after a time, it may be obvious that one topic has generated the most interest, or has the most positive support.

- Creating new team norms. It might be time to add an indecision-breaking strategy to the team norms: a simple majority vote, or agreeing to do one topic this cycle and one the next.

Some Challenges in Topic Selection

Even when teams pay attention to these criteria for selecting lesson study topics and engage in a careful and purposeful team decision-making process, many challenges still exist in selecting mathematical topics that teams learn to overcome with experience.

Choosing Nonmathematical Topics

Students have a lot of difficulty with many kinds of skills that aren't specifically mathematical, for example using technology or taking good notes. These skills are certainly important to student learning, but we have learned that teams will get the most out of their lesson study experience if they choose a mathematics topic *and* can state a mathematical understanding goal for the topic. Why? Because the primary question teams are researching is "How does our lesson promote student thinking and understanding of mathematics?" If your team wants to investigate something like the use of a new technology, that is OK, but try to do it within a lesson that has mathematical understanding as its *primary* goal. For example, a team could choose the topic "understanding the laws of exponents" and plan the lesson to build that understanding. As one aspect of the research, the team can plan to observe how student mathematical thinking about exponents is supported or affected by use of the calculator, or what misconceptions the calculator use seemed to promote or clear up. This is quite different than choosing "using the exponent functions of the calculator" as your main topic.

A similar question arises about topics that focus on a type of pedagogy or a process standard. Many teachers are experimenting with a problem-solving approach or inquiry-based instruction, or want to strengthen their students' skills in communication or in using multiple representations. In fact, lesson study is a great way to investigate instructional approaches and process standards. However, the team should still ask: What will our mathematical understanding goals be? What is the content of the lesson? What mathematical thinking will we observe? Is "representation" what we are teaching, or is it something used to build understanding of mathematics—and if so, what mathematics? If your team is leaning toward a topic of this sort, we suggest that you pair it with a mathematical content topic, for example "What role does representation play in student thinking or learning about proportion?"

Finding a Topic of the Appropriate Size

Choosing a topic is often hard, especially for new teams. One snag that teams often hit in making their topic choices is what we call finding the appropriate "grain size"—that is, choosing a topic that is not too big, not too small. Ultimately, the team will choose a topic suitable for a single lesson. Usually the team begins by identifying a "unit-size" topic—something that would require a week or less to teach—or at most two weeks. Then the team sketches out a unit outline, and identifies the day when the research lesson will take place. Getting to that single-day topic from something larger can involve very valuable discussion. The team must begin to unpack the mathematics and refine their thinking about what they want their students to understand, how understanding will build within the unit, and how various topics are connected. To do this kind of unpacking requires stepping back from the way topics are listed in the contents of textbooks or in state and national standards, moving away from a "What do we have to cover?" mind-set, and focusing on how students will best learn the topic.

Negotiating Topics for Cross-grade Teams

Teachers on teams representing multiple grade levels face a few extra hurdles in topic selection. They may wonder: Will the work we do be relevant to everyone on the team? Will we all have expertise to offer on this topic? Who will be able to teach the lesson? In our experience, the cross-grade team often shifts quickly from seeing their diverse membership as a challenge, to seeing it as a benefit. After some discussion, team members often realize that they stand to learn a great deal about the team's broad goals, no matter the grade or class where the lesson will be taught. Also, as soon as the team begins discussing the mathematics of the topic and how students think about it, talking with teachers from prior or later grades is so enlightening that teachers are no longer concerned about the grade in which the lesson will be taught. For teams where the grade difference *is* a real hurdle, the team often settles it with a compromise—that is, this cycle we'll do sixth grade, next cycle seventh grade, or, let's choose a topic that everyone on the team can teach because it connects to multiple grades. Some cross-grades teams chose a mathematical topic that spans multiple grades, and then try to develop two lessons addressing the topic at two different grade levels. (See Chapter 20, *How Lesson Study Changed Our Vision of Good Teaching* and Chapter 22, *The Long-meadow Story: A District Lesson Study Initiative* for examples of vertical teams using lesson study.)

Improving Your Team's Lesson Study Practice in Topic Selection

What does it mean for your team to be doing well in topic selection, or to improve or deepen your practice in this regard? What are some of the key differences in how a novice team and one that has developed a strong sustainable lesson study practice might approach this choice? Topic selection is impacted by a complex set of factors including team composition, school setting, teaching experience, lesson study experience, and so on. Novice teams often select their topic almost entirely based on the timing of the lesson because the other factors haven't been suggested yet. Teachers in their first cycle of lesson study are learning about so many new aspects of the process that just trying the steps of the lesson study process in their most basic form may be the best choice. By contrast, an experienced team might have discovered that they got the most out of topics that they themselves were most frustrated about teaching, or that contained mathematics they didn't understand well. A strong team may have a real sense of what it means to pursue their own research questions about teaching mathematics and over time develop a way to choose topics that support their research focus.

Reflection Questions

The following questions might help you and your team think about how to strengthen your topic selection, or to deepen your learning from lesson study through your choice of topics.

- *Is our selection of topics purposeful?* Are all of our choices driven by timing, or are we considering topics because of their importance to student learning? Have we asked colleagues in the upper grades to suggest areas where students enter with weak understanding? Does our team have a rationale for our choice? Over time, teams get better at choosing topics that matter, *and* in explaining this choice to colleagues in the "rationale" section of their lesson plan or at a post-lesson discussion. Doing this explaining pushes the team's thinking, allows the team to draw better conclusions, and offers readers/observers of the lesson a more coherent vision of what the team has studied and learned.

- *Is our team developing a research theme?* Does our team have a larger, longer-term research theme it wants to pursue deeply, over many cycles? A team can use topic choice to develop or support a long-term research theme or broad goals. For example, a team might have a strong interest in understanding how students think about functions. The team could decide to pursue that through multiple cycles in a systematic way—perhaps exploring students' understanding of both explicit and recursive expressions for each type of function. A team that has pursued such a research theme over time can develop a number of related hypotheses, see these tested in many classrooms, and have a lot to report to the field—perhaps in the form of a published article or compilation.

- *Does our topic selection provide us with opportunities to learn?* Is our team articulating learning goals for our team, in addition to our learning goals for students? Have we chosen topics that provide us with opportunities to learn in both of these arenas? Do we sometimes choose topics in areas where our content knowledge is weak?

- *Is our team functioning well as a professional community?* Does our team share knowledge and ideas, critique each other's ideas, make group decisions effectively, and avoid territorial battles in our topic selection process? Has our choice of topics helped our team address issues important to our colleagues and to our school community? In choosing topics, are all viewpoints on the team respected?

| # Topic Study: Understanding the Topic and How It Is Learned

What's the first thing most novice teams want to do once they've selected a topic for their research lesson? Plan the lesson, of course! In some settings that is the natural next step. However, in lesson study the next step is slightly different—teams investigate textbooks, standards, scope and sequence documents, and other mathematics resources *before* planning the lesson. This research gives the team a stronger foundation of knowledge about mathematics, and about how students think about and learn that mathematics, on which to develop their own lesson. It also helps them locate high-quality existing lessons or problems from which to build. The process goes by many names: topic study, topic research, or text analysis. The Japanese call this process *kyouzai kenkyuu,* which translates literally as *investigation of instructional materials* and is seen as an important way for teachers to improve their practice not only for this one lesson, but throughout their careers.

In the Lesson Study Communities in Secondary Mathematics project, topic study was initially difficult for us. We, and the lesson study teams we coached, had never done anything quite like it. Perhaps we also did not fully appreciate its function and importance in the lesson study process. Over a period of many years, our successes and failures with topic study led to a better understanding and appreciation. In this chapter, we share some of that hard-won knowledge to smooth the path for other leaders and teams. We encourage all readers to be patient; it's an area of lesson study practice that's worth pursuing throughout your career. The purposes of topic study, what activities lesson study teams engage in that help them understand their topic, and how the knowledge the team gains is used and shared are discussed.

Why Is Topic Study an Important Component of Lesson Study?

In *The Teaching Gap*, Stigler and Hiebert point out that "the goal [of lesson study] is not only to produce an effective lesson but also to understand why and how the lesson works to promote understanding among students" (Stigler and Hiebert 1999, 113). Topic study helps teams develop this kind of knowledge. Questions teachers learn to ask during topic study eventually become questions they naturally consider as a regular part of their teaching practice. Consider this reflection from a lesson study participant:

The process used in lesson study of continuous questioning, reflection, analyzing and reworking is invaluable. There is not one lesson plan, unit activity or syllabus that I work on that is not impacted by my experiences with lesson study. I have changed entire projects and orders of tasks learned by students due to my involvement in lesson study. Each project I design and complete with students now goes through the same questioning and development stage that we modeled in the lesson study team. Researching and documenting the supporting academic concepts involved in each of the learning units taught has raised the expectations I have for my students.

What Are the Main Goals of Topic Study?

The team's aim is to understand the mathematics of the lesson and unit and to clarify how students think about and come to understand this mathematics. Teams address a set of key questions to refine their student learning goals and develop hypotheses to test through their research lesson.

- What is the full mathematical context of this topic?

- How do students think about and learn about this topic?

- What prior knowledge will students bring to the lesson?

- What mathematical ideas do we want students to understand in the lesson?

- What pedagogical approaches might help students build this understanding?

- What will this lesson prepare students to learn in the future lessons and years?

This topic research is often captured in a written summary of the rationale for the lesson. This summary includes the mathematical context of the lesson, and the team's ideas for how students learn this mathematics. The process of broadening knowledge through topic study, then using this broad knowledge to craft a lesson on one idea, is epitomized in a Japanese teachers' saying:

To teach one you need to learn ten.
But remember, you are only teaching one!
So, after learning the ten you must discard nine.[1]

How Do Teams Study Their Topic?

There are several strategies teams use to study their topic and broaden their knowledge of the mathematical context of that topic. These strategies include using the team's expertise as a starting point, doing mathematics together with colleagues, investigating instructional materials and other resources, and seeking outside expertise.

---

1   Watanbe (2007). Dr. Tad Watanabe at the Chicago Lesson Study Group Conference, May 2007, used this saying to explain *kyouzai kenkyuu*, the Japanese term for investigating instructional materials.

Use the Team's Expertise as a Starting Point

A natural and productive way to begin mapping out the mathematical territory of the lesson is by sharing what team members know from their prior teaching experience and other studies. This establishes common team understandings and standardizes the language about the topic. Sometimes we don't even know what we know, until we step back, think about our years of classroom teaching, and formulate explicit hypotheses about how students might learn the topic.

These discussions are often a very energizing experience for lesson study teams. Teachers share instructional approaches that they have found effective in teaching the topic, or places where students typically get stuck in learning about the topic, and they develop theories about how students learn the topic. They may debate what prior knowledge is needed for students to understand a topic, or even whether you can teach a concept without all the ideal prerequisite skills in place. Comments like the following might be heard during these discussions:

> *I've always thought similarity should be taught before congruence because . . .*

> *My students think of the y-intercept as a number (b), but I notice that the textbook refers to it as a point . . . which do you emphasize in your classes?*

> *I've always known how to divide fractions, but I'm not sure how to explain to my students why the invert and multiply algorithm works.*

> *Why do our students have so much trouble understanding area and so often confuse it with perimeter?*

In a more extended example, one of the middle school teams we worked with developed a seventh-grade research lesson on comparing the volumes of cones, spheres, and cylinders. The team had chosen to focus on volume because it had recently been added as a seventh-grade topic. They began their topic research by sharing their prior teaching experiences and discussing their expectations of student understanding in the seventh grade:

- When our students were in elementary school they experienced an informal approach to solid shapes—mostly volume as *contents* or *filling*. Then in fifth or sixth grade, students calculate volume of rectangular prisms with the formula volume = length × width × height and the prior idea of volume as filling gets lost. The filling idea is very basic, but we don't want students to lose that concept. How can these two ideas be better connected? What level (or type) of understanding volume seems appropriate for seventh grade?

- What other ideas are students encountering in seventh grade that might connect well with their study of volume? What do students need to know to be prepared for studying 3-D shapes in high school geometry?

- What are students' understandings, basic assumptions, and misconceptions of volume? What does it mean that they confuse surface area with volume?[2]

This sharing of expertise was the team's first step in topic study, and helped them to unpack how their students learn about volume. It also raised a lot of questions about students' development of ideas on volume across the grades, in particular, when (and how) volume *formulas* are best introduced. To help them answer these questions the team reviewed the relevant state mathematics standards and problems in their textbooks, noting that in both sources, seventh grade seemed to be an inactive period in the learning trajectory on volume.

In many cases, including this one, teams don't consult research on children's mathematical thinking about the topic—perhaps because it is hard to find, or simply doesn't come to mind as something to do. Had this team consulted developmental research by Piaget on conservation of volume it might have provided some insight into how students' ideas about volume develop through the years, and raised some new questions about teaching the topic. In our journey as coaches, we have become more and more likely to encourage teams to seek fresh perspectives and new knowledge in this way. New toolkits[3] of resources for topic study are becoming available that include research readings on a topic, as well as lessons and textbook treatments.

Doing Mathematics Together with Colleagues

For all teams, doing the mathematics problems that are candidates for the research lesson is a powerful strategy for coming to understand the topic mathematics *and* student thinking. This means solving the problems, comparing approaches, generating alternate methods and extensions, considering causes of common student errors. It is sometimes tempting to skip doing the problems, simply because one solution method may come quickly to mind. But *doing* the mathematics together will surface any questions the team might have about the mathematics involved, and it will add to their ability to imagine the problem's role in a lesson. Purposefully generating multiple approaches, assessing the cognitive demand level of the problems, and trying out suggested extensions will leave the team a richer picture of the mathematics and help to build the team as a mathematical community.

One middle school team that was focusing on linear equations (in particular the difference between equations of the form $y = kx$ and $y = kx + c$) worked together on the following problem.

- A big party is being planned and everyone will sit at hexagon-shaped tables. Many tables will be pushed together to make one long table. If 57 tables are

---

2   Team discussion summarized from Jane Gorman's coaching notes of team meetings.

3   Mills College Lesson Study Group researchers have toolkits on fractions, proportional reasoning, and area under development. See www.lessonresearch.net for information.

pushed together, how many people could sit at the table? Keep in mind that tables are joined at a side (edge) not at a vertex and that only one person can be seated at a side (edge) of the hexagonal table.

- Find a way to accurately predict how many people could be seated, given any number of tables?

By working on the problem, team members discovered together quite a variety of rules that could express this relationship,[4] discussed the difficulties students have understanding the difference between directly proportional relationships and those that are linear but not proportional, and generated a series of extension problems like "What happens if the shape is a pentagon?" "What happens if the tables can be joined in any way, not just in a line?" They also debated whether the y-intercept was meaningful in this context.

A team seeking to deepen their understanding might also explore problems that stretch their thinking *beyond the intended grade level*. This exploration is intended to help the team understand what thinking is required by the problems, what prior and later learning is related to the problems, and how students might approach them. Collaborative exploration of the mathematics through problems helps the team to see how the key mathematical concept relates to the learning trajectory for a mathematical idea, to see connections across mathematical topics, and to see connections across grades, thus placing the topic in a broader mathematical context.

Investigate Instructional Materials and Other Resources

Investigating "the raw materials of teaching" (e.g., textbooks, lessons, standards, teachers' guides, curriculum scope and sequence, Internet resources) is the heart of topic study, and helps the team articulate team hypotheses about how students learn the topic (Watanabe 2007). By studying these materials, the team gains a more nuanced vision of what the student learning progression and challenges might be. They will examine their instructional materials, focusing on the order of the topics in the textbook, the expected content and level of understanding at different grade levels or within different units in a given grade, and the pedagogical strategies recommended for students at different developmental levels. The team will consider why the textbook materials might present the topic in the way that they do, and why different textbooks differ. The team should consider how each approach might support or inhibit independent student learning of the concept as well as the teacher's role in each approach. Catherine Lewis has commented that "lesson study is most productive when educators build on the best existing lessons or approaches, rather than reinventing the wheel . . . if your group searches out and studies the best existing lessons, it will result in a better

---

4   Rules generated included: Rule 1: $4t + 2 = s$; Rule 2: $6 + 4(t - 1) = s$; Rule 3: $2 \times 5 + 4(t - 2) = s$; Rule 4: $5t - (t - 2) = s$. (t represents the number of tables and s represents the number of seats.)

research lesson and help create a system that learns rather than one in which every group of educators reinvents the wheel" (Lewis 2002a, 62–63).

For example, one team that chose *slope* as the focus for their eighth-grade research lesson decided to have each team member review one textbook, then discuss the varied approaches at a team meeting. The team summarized their findings:

> *Our research identified two major methods of instruction. Our textbook and most of the others we reviewed followed a traditional approach with students moving from fractions to ratios [and] proportions, rates, functions, graphic representations of functions (usually proportional relationships/direct variations), then slope. [In these texts, slope is approached within the unit on graphing lines from the* y – mx + b *form, and students calculate slope as change in* y *over change in* x, *or rise over run.] Problems and investigations were interspersed throughout each topic. However, connections from one skill/topic to the next were primarily inferential with the teacher being responsible for interpreting and delivering bridge activities or conversations. This approach is more "recipe-like" in its presentation of concepts, often as discrete skills, then a series of drill/practice examples, and applications within problem-solving situations.*

> *The less traditional approach, typified by CMP,[5] is organized by thematic units with a series of investigations that introduce the concept of slope without actually identifying it by name. Students come with a background "threading" fractions to ratio and proportion, to "rate of change" instead of the term slope. Similarly Singapore Math engages eighth-grade students in manipulations before addressing the concept of slope. This approach is often identified by an introductory problem-solving situation where skills and concepts are embedded within the problem and "discovered" as students work through toward a solution.[6]*

Seeing that the less traditional approach was more based on problem solving, and also aimed at conceptual understanding over calculation, the team decided to pursue the latter, problem-solving approach.

To begin topic study, a team should carefully examine their own curriculum and texts, and state and national standards.[7] The standards will outline a basic set of performance standards, indicating a rough framework within which the topic sits, but studying textbooks provides a great deal more. Comparing two or more texts with markedly different approaches can be very informative. Discuss, "What does each approach imply about how students learn this topic?" and "What are the mathematical

---

5   The team had explored the topic in the *Connected Mathematics Program* (CMP) textbooks (Lappan et al. 2006).

6   Lesson background information, Ahern Middle School Open House Lesson, Foxborough, MA, 2004. Lesson prepared by cross-district team from the Lesson Study Communities in Secondary Mathematics project working with Dr. Akihiko Takashi.

7   These include the *Principles and Standards for School Mathematics* (NCTM 2000) and *Curriculum Focal Points* (NCTM 2006). See also Resources Appendix for additional mathematical resources.

differences in the approaches?" The text and teachers' resources sometimes include the following components to help identify the key mathematical ideas:

- *Detailed explanations of the mathematical content of the topic*: What are the major concepts students need to understand?

- *Pacing guides, unit outlines, lesson plans, and unit assessments*: In what order should the concepts be encountered to build understanding? How are the ideas presented to students?

- *Problems for extended investigation*: What problems engage students in thinking about the topic? What problems require students to engage in problem solving?

- *Suggestions for pedagogy:* What are key questions to ask in teaching the lesson?

One important outcome of studying the textbooks and standards will be a description of the possible mathematical learning trajectories for the topic—how students think about and learn this topic over time. This could be something similar to a concept map, a narrative explanation, or a chart showing the learning goals at different grade levels. Fielding student questions during a lesson and deciding the order in which students should present their solutions can be much more purposeful and effective when you have this larger vision of the mathematical territory and expected learning trajectories. Having a vision of this trajectory then helps the team state and refine their understanding goals for the lesson.

Seek Outside Expertise

The role of topic study in lesson study is extending the team's knowledge. Another way to do that is seeking additional knowledge from a content expert or through existing research. Some teams invite a university mathematician or mathematics education specialist to join the team. Others select a thought-provoking article as a source for ideas for the lesson content or pedagogy. A team could even decide that taking a summer course or workshop on their content focus area would provide them with a needed boost in background knowledge. (See Chapter 16, *Incorporating Expertise from Outside the Team*.)

One of the lesson study teams with which we worked was participating in an NSF-funded Mathematics and Science Partnership program that brought together university mathematicians and study groups of secondary mathematics teachers. Because this school was already actively engaged in lesson study, the university mathematician joined the lesson study team. In one of their lesson study cycles, the teachers identified combinatorics as a topic that many of them were interested in learning more about for themselves, as well as for teaching their students. The university mathematics professor supported this team's learning by offering problems for the group to work on together and by making connections between the ideas about combinations, permutations, and the patterns in Pascal's triangle. This idea of drawing on an outside expert, sometimes referred to as a "knowledgeable other," is common in the Japanese practice of lesson study.

An additional resource to which teachers have access is their students. Lesson study teams sometimes choose to try out problems or parts of problems related to their lesson topic with classes of students prior to the formal observation in order to gather some data about the development of student understanding of the topic. Some teams develop preassessments to administer to students and/or examine student assessment data to learn more about students' current understanding of the topic.

Summarize Research—Frame Understanding Goals and Rationale

Following this period of knowledge gathering and learning about the mathematics of the team's topic, it is important to capture what has been learned about the topic in some written form. Some teams prefer to summarize their learning in a narrative rationale for the lesson. For other teams, a concept map or standards table may better present the mathematical context and learning trajectory; an added paragraph explains how the team's approach in the current lesson is related to the mathematical context and learning trajectory.

The importance of putting this learned knowledge and rationale into writing may not be obvious to a new team but has several benefits. These benefits include:

- Helping the team clearly articulate their hypotheses about student learning.

- Informing observers of the research lesson about the mathematical context and the team's theories, so that those observers can contribute to the data collection and discussions.

- Sharing the team's findings with the profession. Research lesson reports that include the team goals, instructional plan and rationale, notes on the post-lesson discussion, and team findings are a significant resource to other teachers and researchers.

Thinking on a scale that includes the full profession, the team's learning is a contribution to building a larger professional knowledge base about teaching. We can imagine building a written repository of research lessons, similar to that which is available in Japan. There, bookstores have whole shelves filled with research lesson reports that have been written by teachers. These reports capture the teachers' developing knowledge for use by other teachers. Teams include with their research lesson plan a lesson rationale section, where they can describe the hypotheses they had and how those hypotheses and what they learned about the topic led them to design the lesson as they did.

Conclusion

Learning how to do topic study can be challenging, in particular for novice teams who are not accustomed to analyzing instructional materials or seeking out additional expertise to inform their teaching in these ways. Be patient. Even a modest exploration of the topic mathematics with your team and a discussion about how the topic is taught in different grades can have a powerful impact on the research lesson, and over time your team will find ways to deepen your learning during this phase of the cycle.

Being reflective and persistent also matter. Questions that teams and leaders might ask themselves are:

- If our lesson went poorly, could it have been because we didn't study our topic deeply enough? (Going back to the Japanese saying . . . Did we "learn ten"?)

- If we collected lots of information, but we aren't sure what is most important to focus on for our lesson design or rationale . . . should we revisit the key questions we started our topic study with, for example: What do we want students understand? What does it mean to understand this topic? How do students think about this topic? (Returning to the other half of the saying, "Are we at 'one' yet?")

- If our lesson was great but we aren't sure what importance it has for other teachers, did we seek out current knowledge in the field (lesson plans, research, etc.) as a starting point?

- If reflecting on our learning at the end of the cycle seems very distant from our topic study. . . . Did we record our topic study findings, so we could connect it to our observations of students and to our team learning?

Ongoing reflection about improving your learning through topic study should, over time, deepen your knowledge of mathematics, teaching, and student learning, and lodge the team's work in a growing body of professional knowledge.

| # Develop the Research Lesson

P hase 2 is about envisioning the design of your team's research lesson. The lesson design must support the goals and research agenda that your team set for itself at the beginning of your work together. Developing a research lesson includes selecting problems or tasks, determining the pedagogical approach, and preparing all necessary handouts, manipulatives, and so on. By debating various approaches and coming to consensus on what to try in the research lesson, your team is developing a theory about how students will come to understand this mathematics. Through all this discussion, your team will become solidly grounded in the content and the lesson. The research lesson should end up being a team lesson based on everyone's ideas (i.e., no individual should think of the lesson as my lesson based on my theories). The full mathematical context and implications of the lesson become clearer during this phase. The final product of this phase is a detailed lesson plan and rationale for the lesson design that can serve as a teaching tool, a research tool, and a tool for disseminating the team's learning. The detailed lesson plan and rationale does not need to be perfect, it just needs to be good enough for the team to test out their ideas and gather some data—there will be more time for revising the lesson during Phase 3 of the lesson study cycle.

This section of the *Leader's Guide* contains chapters that will help your team picture how the team will design their research lesson, how a specific part of the lesson plan—anticipating student responses—provides powerful learning for the team, and how the team will incorporate all of their decisions into a written and multipurpose research lesson plan. The set of activities and questions that are central to the team's work in Phase 2 are described in the chart, followed by notes about chapters for further reading.

Key Activities	Central Questions	Related *Leader's Guide* Chapters
Develop the research lesson and make a detailed written version	*During lesson development the team considers all of the questions in this column.*	Chapter 6, *What Makes a Good Research Lesson?* Chapter 8, *Anticipating Student Responses* Chapter 7, *Developing the Lesson Plan*

Key Activities	Central Questions	Related *Leader's Guide* Chapters
Discuss how the lesson incorporates research theme and goals	• *How will the lesson further our broad goals?* • *How will the lesson engage and motivate students?*	Chapter 6, *What Makes a Good Research Lesson?* (Also, Chapter 3, *Goals and Research Themes*)
Explore/study mathematics and consider the development of mathematical ideas	• *What difficulties do students have with this topic and what do they already know about it?* • *What concepts are key to developing understanding of this topic?* • *How will the lesson help students develop understanding?*	Chapter 6, *What Makes a Good Research Lesson?* Chapter 7, *Developing the Lesson Plan* (Also Chapter 5, *Topic Study*)
Anticipate student responses	• *What student responses/ methods are anticipated?*	Chapter 8, *Anticipating Student Responses*
Anticipate teacher's questioning, lesson structure, and pedagogical strategies	• *What are the key questions the teacher should ask?*	Chapter 7, *Developing the Lesson Plan*
Prepare for the observation	*See Phase 3.*	*Various chapters in Phase 3*

▌ Leadership Focus in Phase 2

During Phase 2, leaders working with teams should allow space for the team to take the lead on developing the lesson, engaging in the activities and questions described. A few areas of focus for leaders during this phase as a whole include:

- Asking questions to focus team members on mathematics and on their goals

- Assisting the team in developing team ownership of the lesson

- Offering outside perspective or expertise, if needed

- Encouraging teachers to do mathematics and to maintain a research stance

- Fostering open discussion and listening to others' ideas

- Facilitating discussions, if needed

CHAPTER 6 | **What Makes a Good Research Lesson?**

To teach one you need to learn ten.
But remember, you are only teaching one!
So, after learning the ten you must discard nine.[1]

The moment when a team starts to develop a concrete plan for the research lesson is a major turning point in the lesson study cycle. Phase 1 was about building broad knowledge of the mathematics. Developing the lesson is about focusing on specifically how students will think about and come to understand the mathematics during this one lesson. Phase 1 discussed setting goals. Developing the lesson is about using these goals as guiding principles in lesson design and asking, "How does this problem, this manipulative, or this pedagogical strategy support our goals?" In the folk wisdom of the quote, Phase 1 was clearly about *learning ten*. Developing the lesson is clearly about *teaching one*.

If you have been doing lesson study for a while and think for a moment, "Which were our best research lessons?" there is a good chance that what comes to mind are those where something really interesting or surprising happened, where you had a new insight about teaching, or where the team had spirited debate at the post-lesson discussion. One might have been a lesson that went wrong in some way that taught you something important. A good research lesson doesn't have to be flashy, foolproof, or highly innovative. It is one in which *learning takes place* when you observe and discuss it, but also when you write it, when you revise and reteach it, and when you pull it all together in the end. In short, a good research lesson is one that lets you *and* your students learn. For teachers, it is a chance to test ideas and to get new insights about teaching and learning of mathematics. For students, it should be a strong mathematics lesson, address local student challenges, and build on the best current ideas about teaching mathematics.

How do teams go about envisioning such a lesson? Gaining experience through your own research lessons, and seeing other teams' lessons will help the most. In this chapter, we share what we have learned through our own research lesson experiences, including: thoughts on overall principles teams can keep in mind as they plan their

---

1   Dr. Tad Watanabe at the Chicago Lesson Study Group Conference, May 2007, in his talk on *kyouzai kenkyuu*, the Japanese term for investigating instructional materials.

lessons, features shared by many of the most illuminating research lessons we have seen, and an assortment of instructive examples.

Design Principles

If one had to choose one part of the lesson study process where it is easiest to lose sight of the big picture, it would be when the team is deep in the nitty-gritty of lesson planning. The following basic design principles may help your team see how the big themes underlying the entire cycle (focus on students, mathematics, etc.) play out during planning of the research lesson.

Mathematics Teaching Principle

Your primary goal should be to design a strong mathematics lesson. Reflecting team expertise and current research on effective teaching of mathematics, the lesson should foster problem solving and include features like the following:

- Lesson tasks and activities that ask students to think, reason, and make sense of the mathematics.

- A lesson that encourages multiple approaches to problem solving.

- Problems that are aligned with the mathematical understanding goals.

- A lesson that asks students to communicate about mathematics.

Research Principle

Make the lesson an experiment and include features that support the experimentation:

- The team has ideas, questions, and hypotheses about student learning, and has articulated these in a rationale statement so that observers can partner with the team in this research.

- The lesson is designed with data collection in mind so that post-lesson discussion and analysis can be based on evidence. It is designed to *reveal student thinking* so that observers will be able to see and hear students.

- The lesson has a deliberate evaluation element so that the team is aware of student learning as it progresses during the lesson.

Teacher Learning Principle

Keep teacher learning and team ownership of the ideas as a high priority throughout the lesson development process:

- The lesson relates directly to teachers' daily lessons and challenges. Working on it will generate important discussions, build useful knowledge about common challenges, and generate curiosity about new questions.

- The lesson is part of a larger professional learning agenda—that is, the team or school is pursuing a research theme over time, and/or the team research builds on and contributes to professional knowledge on teaching this topic.

This is not intended as a to-do list, but as a reflection tool. The more that teams gain understanding of lesson study, and think more systematically and long term in choosing a focus for their work, the more these principles come into play. The result is not necessarily flawless lessons, but lessons that have the qualities of a good experiment. Asking questions related to these principles as they design the lesson, teams will begin to "own the ideas" so thoroughly that they can join into a lively post-lesson discussion and openly debate the significance of observation data.

Where to Begin?

There is no single lesson prototype that teams follow in creating their lessons. Many lesson styles are possible. However, experience suggests that certain lesson types easily accommodate these principles—for example, a lesson that begins by engaging students, then has students work on a challenging problem, and culminates with a discussion of the mathematics of various solutions. Some types are more problematic because they ignore one or more of the design principles. For example, in a lecture students' thinking may not be revealed. Generally, teams feel free to try new kinds of lessons as well as experimenting with recognized strong pedagogical models.

Which Lesson in the Unit Is Our Research Lesson?

Teams ordinarily begin by sketching out a unit on their topic and deciding which point within that unit is most critical for student learning and most revealing of student thinking in ways that will inform the team. Hypothesizing where that *teachable moment in the unit* is represents the first step toward developing a good research lesson.

Arch Lessons

Often teams have chosen a topic that students have great difficulty learning or remembering. They know that even when teachers have taught the topic well, there are important conceptual understandings that even a conscientious student would still be lacking. Many teams facing this challenge decide to create a special "extra" lesson—one that is aimed at building solid understanding of the elusive big ideas and placed at a critical moment in the unit—often at the very beginning. One lesson study coach called these lessons *arch lessons* because the team envisioned the unit as a developing sequence of ideas, which, like stones in an arch, would collapse if one important stone were missing. Locating where this missing stone belonged and finding a new way to convey those ideas were the focus of the team's research.

One of these arch lessons came from a team who thought that the students emerged from Algebra 2 with too narrow an understanding of logarithms. The textbook unit defined exponents, then developed their properties for integers, rational, and real numbers, and ended by introducing logarithms as solutions to certain exponential equations (that is, as inverses of exponential functions). Five decades ago, before handheld calculators, logarithms were objects that were more basic, more concretely useful, and much more concrete. One used them to calculate, and developed understanding of their functioning and nature by working with them (by hand!). This team wanted their class to see these numbers in action again.

The team developed a lesson featuring applications that use logarithmic scales: the Richter scale for earthquakes, the Scoville scale for classifying the heat of a red pepper, and the decibel scale for measuring the loudness of a sound. Students computed with these numbers, and constructed a logarithmic ruler, diagrams, and pictures to illustrate the scales. These tasks instilled the logarithmic habit of thinking into the daily vocabulary of the students. Throughout the entire lesson there was no use of the term *logarithm*. The notion of *orders of magnitude* was easy to introduce into the discussions, and this proved to be a helpful backbone for the informal discussions and relative calculations. One team member remembers this as their most powerful lesson:

> For me the hot peppers lesson was the most powerful for a couple of reasons. One reason is that it really helped me appreciate the properties of logs. The lesson can be used and adapted to meet the needs of students of varying ability. I think the students really got the concept of a log. The concept was presented in a new setting and put into context.

What made these arch lessons successful as research lessons? The team members thought hard about exactly what understandings they were after, how these understandings would develop across a unit, and where a big idea lesson would have the most impact. Clearly, the lessons incorporated the design principles (challenging mathematics problems designed to reveal student thinking) and also found ways to put students in a position of considering ideas in a very fresh context, often concrete or realistic, that generated high interest and active mathematical thinking. In many cases, the teams also made a conscious decision to launch the unit with the lesson and refrain from reviewing or introducing technical vocabulary and algorithms until later lessons. This forced students to reveal prior understandings, and to come up with their own ways of doing the problems. For an extended example of this type of lesson, see Chapter 19, *Josh and Betty's Lesson Study Journal*, where the authors describe how they developed a launch lesson on triangle congruence.

Good Problems—An Important Ingredient
Many of the best research lessons we have seen center the whole lesson around a few good problems that directly address the team's student understanding goals and call for higher-order thinking—analysis, sense making, deduction, problem solving, generalization, finding patterns, organizing. In thinking about the mathematics teaching and

research principles, this type of problem contributes to the research lesson design by calling for *and* revealing student thinking, and providing opportunities for students to problem solve, use multiple approaches and representations, and talk together about what they are thinking. Often they contribute also to the teacher learning principle, as the team has to figure the problems out! We have learned from our Japanese lesson study advisors that the word *hatsumon* means "asking a key question that provokes student thinking at a particular point in the lesson" (Shimizu 1999, 109). Although the meaning refers to good teacher questioning throughout the lesson, many of us focused on questions (or problems) that might center a whole lesson. Some of the following examples are classic problems; some started as a simple problem from the text; some the team made up. All combine accessibility with challenge and openness to multiple solution methods.

- *If you have a cone and a cylinder and a sphere all with the same radius and height, how would you order them by volume smallest to largest? Where would the hemisphere fit in?* This launched a seventh-grade class (who knew no conic volume formulas) in a hands-on investigation to deepen students' concept of volume-as-filling before formulas are introduced, widen their physical experience with geometric solids, and increase their ability to use ratios and percents in comparison statements.

- *Let's calculate 64 × 23 and 46 × 32. What do you think about the products of these two problems?* This problem, from a sixth-grade research lesson conducted at the 2001 International Congress of Mathematics Educators in Japan, encourages students to think about the structure of computation as they explain why the two products are equal and consider whether other pairs of products will have this same property.[2]

- *How many handshakes will there be if all twenty-five students in the room shake hands?* This problem was the centerpiece of a lesson aimed at increasing students' ability to generate and compare multiple solution strategies in a problem-solving situation and is described further in Chapter 20, *How Lesson Study Changed Our Vision of Good Teaching.*

- *What are the congruence theorems for quadrilaterals?* This problem was chosen to give students in an honors geometry class experience in developing theorems, proofs, and counterexamples and in pursuing systematic approaches to exploring cases.

2   *The Mystery of Calculations*, research lesson taught by Masami Fukushima (January 1, 2001) at the Elementary School attached to University of Saitama, *School Mathematics in Japan*, Mathematics Education Division, Institute of Education, University of Tsukuba, Japan,

- *What quadrilaterals will tile the plane?* This problem was chosen to connect students' mathematical knowledge to practical skills such as utilizing wood efficiently in construction. The lesson is described in more detail in *The Essence of a Day*.

- *Does a parabola have slope?* This problem was used to transition from a unit on linear functions into a unit on quadratics to encourage students to think about the pattern of growth for quadratics, using more than one representation. Students tried to relate the change in *y* and the change in *x* as they had for lines, for parabolic graphs and tables, and conjectured whether the term *slope* had meaning for parabola.

We don't want to give the impression that a great problem equals a great research lesson, or that the problems have to be big ones that could occupy an entire lesson. Whatever the size, the team thinks carefully about the role each problem or task plays in the lesson, how it is worded to encourage mathematical thinking, and how it addresses the understanding goals. The team also asks, "Do we need *all* of these problems?" "Are the problems engaging?" Most importantly, the team will think about teacher questioning throughout the lesson, not trying to produce a script for the teacher, but to include in the plan key questions the teacher might ask to foster learning or discussion.

It is our experience that choice of problems is a major factor in producing a good research lesson. But how the problem will be used within the lesson is equally important: how it is introduced, how students are encouraged to engage with the mathematics, and how it is discussed. The expected levels of cognitive demand and student thinking can easily be reduced[3] if the teacher and the team do not think about the problem in the context of the full lesson and help the teacher anticipate both student responses and the teacher's role. These factors are treated in the next chapter.

Instructive Examples

In our work as lesson study coaches, we are always trying to understand *why* we learn more from some research lessons than from others. Placement within the unit, excellent problems, and strategic teacher questioning are important. However, research lessons that produced powerful team learning (or the lack of) have appeared in many guises. The following examples emerged in our quest, and may serve as fodder for discussion by teams that are thinking about ways to develop good or better research lessons. It would be instructive to consider, for example, how the design principles apply in each example.

---

3 For discussion of factors that reduce cognitive demand levels in the enacted lesson, see Stein et al. (2000).

Exploring New Pedagogies

In one urban district a new, standards-based middle school curriculum had been adopted. The problem-centered pedagogy was foreign to most of the teachers. Teams from five of the schools used lesson study to learn about the new curriculum. Their work focused on collaboratively making sense of the mathematics and imagining what the curriculum's pedagogical model should look like in their classrooms. They considered questions like: What is my role as teacher? How do students learn in this investigatory model? Unlike the team-created logarithm lesson, these research lessons emerged directly from lessons in the curriculum. Because the team-learning goal was to understand how the new curriculum worked, adaptations made in the text lessons were small. Still, the process helped teachers to understand the new pedagogy, and to test their own ideas about many things—for example, how to help students share their solutions after the exploration. The teams altered lesson details to reflect local student needs, culture, and knowledge. In one cycle, all five teams studied the *same* textbook lesson, yet produced five very different research lessons! Each team had focused on a different question about student learning, or a different part of the lesson, and all teams had launched the lesson in a unique way to build student engagement.

This example represents another very viable approach to designing a research lesson: begin with a strong existing lesson or unit, study it closely to understand how it is intended to promote student learning, and make adaptations that address the team's research questions and goals for students. In this case, the approach offered strong support for teachers exploring a new pedagogy.

Worksheets: Friend or Foe?

An excellent lesson study video (Lewis 2005), produced by the Mills College Lesson Study Group, shows a research lesson for fourth graders in which students look for a pattern as they line up triangular tiles. Students recorded their data on a worksheet chart, and most quickly noticed a simple "plus one" pattern in the numbers. Few saw any other patterns, nor were any students able to explain how the pattern related to the tables or create a rule for the pattern. The chart had, in effect, narrowed the mathematical impact of the problem by visually organizing the information in a certain way.

In a Lesson Study Communities in Secondary Mathematics algebra research lesson[4] comparing exponential and linear growth, students began by graphing both kinds of functions and completing a set of written reflection questions about the differences. At the first teaching, students spent thirty minutes on these worksheet tasks, with almost no dialogue. What the team had hoped would spark collaboration in this inclusion class had isolated students in silent work. For the second teaching, the team shortened the graphing tasks and changed the reflection questions to spark more discussion in student pairs.

---

4   This lesson is treated in depth as a Case Study in *Lesson Study in Practice* (Gorman et al. 2010).

Designing handouts is always a challenge. For a research lesson, student written work can provide excellent data on student thinking and understanding, or it can squelch mathematical conversations. Representations used on the worksheet may determine a particular solution method, keep certain methods, misconceptions, or errors from surfacing, or encourage multiple options. This is not to say that worksheets should be excluded from research lessons. Rather, teams should carefully consider what role student written work is playing in team research and student learning, *and be sure to discuss evidence on this topic at the post-lesson discussion.* Given how often teams use handouts in their daily lessons, experimenting with ways to make student written work a positive contribution to the lesson can produce learning for the team with long-term impact on instruction.

Novelty Lessons: The Math Walk

Many teams see lesson study as an excellent way to experiment with a totally new lesson type, and rightly so. In this spirit, several Lesson Study Communities in Secondary Mathematics teams experimented with designs for a "math walk" in which students went out into the neighborhood near the school and did mathematics problems along the way. How tall is this tree? What is your walking speed? Does the weight of the person affect the motion of a playground swing? The goal was to engage students in real-life connections to mathematics, and to increase student enthusiasm for math. In fact, students were highly motivated and appreciated the chance to do something so different from their usual lessons, and the teams learned many things about student engagement and planning a field-trip-type lesson. But these lessons fell short as research on student thinking. Why?

- The lessons were difficult to observe effectively.

- Problem contexts came from the neighborhood and covered many topics. This meant the understanding goals were not very focused.

- Much of the lesson focused on collecting data, leaving the actual doing and discussion of the problems to the next day.

- Students were focused on the instructions—where to go, how to keep their notes, and so on.

All of these issues converged in the math walks to take the focus away from doing (and observing) mathematics. These same issues can arise in the design of far less exotic lessons, and reduce their power as research lessons. For example, in deciding to include a new game or new technology in the lesson design, the team will encounter a similar design challenge—how to keep students' focus on the instructions from putting the mathematics into the background.

Conclusion

In this chapter, we have shared some ideas that may help your team create its lesson design, and some principles that your team can keep in mind as the development process proceeds. By thinking through how this lesson fits into the learning trajectory of the unit, choosing a few good problems, and talking together about the overall pedagogical model your team wishes to try, you will have envisioned the main structure of the lesson—or at least found the main ingredients. The next two chapters discuss how the team builds a focus on student thinking into the lesson, and creates a detailed instructional plan.

CHAPTER 7 | Developing the Lesson Plan

As the team studies the topic, decides on key activities and pedagogical strategies, and anticipates how the lesson will elicit thinking from their students, team members will document all of these decisions in a written lesson plan. This is not just any lesson plan—it is a *research lesson* plan that serves multiple roles for the team.

The lesson plan *is a teaching tool* because it articulates the flow of activities and questions the teacher should follow. It is also a teaching tool because it arms the teacher with a deep understanding of the rationale for the lesson design—an understanding that can help the teacher make informed choices when faced with the on-the-spot decisions that will inevitably arise. The lesson plan is *a research tool* because it identifies the team's hypotheses and theories; it provides notes to guide the observers who collect data about those hypotheses during the lesson observation; and it ultimately summarizes the findings that the team has learned about. And finally, when the team puts into writing their theories, their lesson design, and their learning, it helps the team to solidify their own learning, and also to turn the lesson plan into *a dissemination tool*. The written plan allows the team to share their learning and lesson design (i.e., the research tool and the teaching tool) with colleagues.

Experienced lesson study teams typically include the following three elements in their plan in order to make it useful in supporting teaching, research, and dissemination:

1. *Background Information.* The background information includes the context for the team and lesson, the goals for the lesson, and the team's rationale about how the lesson design supports student learning.

2. *Detailed Instructional Plan.* The plan provides a concise outline of what will happen in the lesson and helpful information for the teacher and observers, as shown in this commonly used four-column[1] style.

Lesson Outline Problems/Activities	Expected Student Responses	Notes for the Teacher	Evaluation of Student Learning
A simple outline of what will happen during the lesson, including key problems	Problem solutions, approaches, errors, misconceptions, engagement, etc.	Reminders on pedagogy, questions to foster learning, practical hints	What observers (and the teacher) should look for as evidence of learning and progress on goals

---

1   For a less complex look, many teams use two columns—one for the lesson activities and student responses, the other for notes to the teacher.

3. *Summary of the Team's Findings.* Not during the lesson planning, but later in the cycle, it is important for the team to incorporate a brief summary of their findings into the write-up of the lesson plan in order to make sense of the research and prepare to share those findings.

We examine the purpose and contents of each of these three elements of the lesson plan using a research lesson, the Rumor Lesson, developed by a team[2] of teachers in the Lesson Study Communities in Secondary Mathematics project.

1. Background Information

The background on the research lesson plan is for *teaching* because it provides justification for choices that will be made while teaching. It represents the *research* component of the lesson plan because it contains the team's hypotheses and theories. And it is useful in *dissemination* because it provides readers access to the full experiment (the context for which the lesson was developed, the instructional plan, the team's thinking and findings) in a form that could be used by other teachers, or by other researchers and lesson study teams who want to investigate similar questions.

Context and Goals for the Lesson

The lesson plan should include enough information about the students and school to help observers visualize the full context of the lesson. In addition, the team sketches a broad mathematical context—where the unit and lesson fit into the curriculum scope and sequence. The team's broad goals and content understanding goals complete the picture.

Putting this information on the plan allows observers to fully appreciate what the team is trying to accomplish in the lesson, and why. Furthermore, writing the goals

> **The Rumor Lesson Context and Goals (selections)**
>
> *Goal:* Students will recall learning from middle school about exponents and develop new understandings of exponential growth and properties of exponential functions that help them transition from the study of linear to nonlinear functions.
>
> *Team-Learning Goal:* To improve the vertical alignment of curriculum by understanding how topics are treated in middle and high school.
>
> *Context:* Relevant state standards from eighth, tenth, and twelfth grades were listed on the plan (e.g., Grade 10 Patterns and Algebra Standard, "Solve everyday problems that can be modeled using linear, reciprocal, quadratic, or exponential functions. Apply appropriate tabular, graphical or symbolic methods to the solution. . . .").

---

2   For more about the lesson study team that developed this research lesson plan, see Chapter 20, *How Lesson Study Changed Our Vision of Good Teaching*.

and context statements helps the team solidify their own ideas about what students already know and will learn. For the Rumor Lesson, the context and learning goals convey this cross-grade team's broader purpose of learning how topic understanding progresses from middle to high school.

Rationale for the Lesson: Hypotheses Regarding the Development of Understanding

The rationale is a statement of the team's vision of the mathematical context of the lesson, and their ideas about how the lesson will build understanding and foster the team's broad student goals. The rationale briefly shares the team's reasoning and justifies choices that were made in the lesson design with answers to questions such as:

> *Rumor Lesson Rationale (selections): Prior to ninth grade, students have studied about linear functions and the rules of exponents.* Students learned linear relationships through the CMP unit Moving Straight Ahead. They graphed real-life data, created tables and developed equations from the graphs.... Students have a difficult time transitioning understanding from linear to nonlinear functions, [and] assume all data has a linear relationship. ... Placing this lesson at the beginning of the high school unit on exponential functions would reveal prior knowledge and misconceptions ... and develop the concept of exponential growth.

- How does the lesson build on students' prior mathematical knowledge?

- How does the lesson build understanding? Why did the team choose these problems and this pedagogical approach? How will the lesson address mathematical misconceptions?

- How does the lesson prepare students for future learning about the topic?

▌ 2. The Detailed Instructional Plan

The instructional plan sketches out the lesson activity and the teacher's role for the lesson. It is not a script. Instead, it presents a map or outline of the intended lesson, with useful information for the teacher and observers—expected student thinking, teacher questioning, and embedded assessment. This plan gives readers a clear vision of *what the team intends to happen in the research lesson*. As you read through the contents of the Rumor Lesson Plan, it is easy to picture how it provides support as both a *teaching tool* and as a *research tool*. Please note that this instructional plan is the version the team prepared for the *first teaching* of the lesson.

Column 1: Lesson Outline Problems/Activities

As the team envisions the overall content and flow of the lesson, they fill in column 1. Questions addressed here by the team include: What activities and mathematics problems will the class do during the lesson? How will the time be allotted? What are the basic pedagogical approaches—that is, individual work, full-class discussion, and so on? The resulting outline allows a reader to easily visualize how the lesson will transpire. To make this reading even easier, text is kept to a minimum and boxes and shading can be used to highlight the problems students will be doing. The goal should be to make it as simple to read as possible.

(Note: We strongly recommend that you do problems 1 and 2 from the Rumor Lesson now to get the most out of reading the remainder of this chapter.)

The Rumor Lesson Column 1

This team used a typical strategy for designing their lesson by starting with the main setting (a rumor spreading) and then crafting the wording of questions to address their understanding goals. Problem 1 is a standard and engaging context for exponential growth. The team thought all students would easily engage with problem 1 and get data for the first five days using a variety of approaches. They stated the problem in a way that did not demand use of exponents, hoping to learn what students knew from their prior exposure to exponential growth patterns. Problem 2 was designed specifically to force students to think beyond the simple repeated multiplication of 3s to conceiving the problem in terms of exponents.

Picture what you, as an outside observer, might learn by reading Column 1 prior to the lesson. For example, the progression of problems suggests that the team wants students to predict the nth day of the rumor by graphing the data for Days 1–5, then testing possible function rules with their graphing calculators. What might this indicate about the team's vision of the role of problem 2?

Lesson Outline of Problems and Activities

1. **The teacher poses the main rumor-spreading problem to the class. Students will work individually, then in groups, then share results in the full class. (15 minutes)**

   The Rumor Problem: Tara, a high school junior, wants to ask Billy to the semiformal. Today, Tara tells her 3 best friends her plans. On the next day, her 3 friends each tell 3 of their friends. The rumor continues to spread in this manner. How fast is the rumor spreading? Model how the rumor is spreading over the first 5 days.

2. **After sharing various representations for first five days, students are asked to work in groups to predict Day 9 of the rumor, then discuss predictions and methods as a full class. (15 minutes)**

   The Rumor Problem, Part 2: How many new people will hear the rumor during the ninth day? You must predict for the ninth day without calculating the sixth, seventh, and eighth days.

3. **Predicting for the nth day and working with graphs. (15 minutes)**
   Teacher plots points for the first five days on the class overhead graphing calculator and class looks at graph together. Teacher poses nth day question.

   How many new people will hear the rumor on the nth day? What formula will give you the nth day without your recalculating all the steps along the way?

   Students put their formulas on the board and in their graphing calculators. Teacher discusses one formula at a time with the class —testing out their accuracy and discussing. Test student understanding by asking for nth day formulas for new spreading factors:

   $$(y = 4^x , y = 5^x, \text{ etc. })$$

4. **Extension (15 minutes)**

   When will the whole town know? Tara's town has a population of 16,000. Suppose that Tara tells 5 new people each day but those people do not tell the rumor to anyone. What if Tara tells 5 people on the first day, who each tell 5 new people on the second day, those people each tell 5 on the next day, and so on. On which day will the whole town know? Describe how these two patterns of growth differ.

   ---

5. **Summing Up (10 minutes)**

Column 2: Expected Student Responses

Column 2 of the lesson plan contains the team's expectations about what exactly students might think or do in response to the problems. When filling in Column 2 the team asks: What are typical responses, misconceptions, or errors we expect from students? What unusual responses might occur that reveal some special understanding or present a challenge in teaching the lesson?

(Chapter 8 discusses the reasons for anticipating student responses and the impact of having this information.) Here, we relate what the Rumor Lesson team included in their plan and what we can infer from that about their theories of student learning on exponential growth.

The Rumor Lesson Column 2 Selections[3]

The team expected a wide variety of methods for modeling the five days. They also expected that students would have few tools to deal with the Day 9 prediction, and would need a hint to help them recall prior learning on exponents.

---

3   For reading clarity, we are including only the part of Columns 2, 3 and 4 that relates to Problems 1 and 2 in this lesson.

Lesson Outline of Problems and Activities	Expected Student Responses
1. **The teacher poses the main rumor-spreading problem to the class. Students will work individually, then in groups, then share results. (15 minutes)**	Modeling for five days. Students will use a variety of methods: • Graphing • Multiplying $3 \times 3 \times 3 \ldots$ on calculator • Tree diagram • Drawings • Manipulatives
The Rumor Problem: Tara, a high school junior, wants to ask Billy to the semiformal. Today, Tara tells her 3 best friends her plans. On the next day, her 3 friends each tell 3 of their friends. The rumor continues to spread in this manner. How fast is the rumor spreading? Model how the rumor is spreading over the first 5 days.	Possible student errors: • Summing the numbers • Using a linear or quadratic model
2. **After sharing various representations for first five days, students are asked to work with group to predict Day 9 of the rumor, then discuss predictions and methods as a full class. (15 minutes)**	Day 9 Prediction Students: • may ignore directions and continue multiplying by 3 • will be stuck and need a hint • will use 3^9 to obtain the answer
The Rumor Problem, Part 2: How many new people will hear the rumor during the ninth day? You must predict for the ninth day without calculating the sixth, seventh, and eighth days.	

Column 3: Notes for the Teacher

After (or during) the discussion of lesson design and expected student responses, the team discussions turn to the question: How should the teacher be teaching this lesson? On a daily basis, teachers trust their experience and instincts to guide them in the actual teaching. For purposes of a research lesson, however, the team records on the plan: How might the teacher order the sharing or discussion of student methods during the lesson? What mathematical understanding is the teacher trying to elicit during a discussion? What would make an excellent follow-up question to ask? It is these pedagogical suggestions, along with reminders, that go in Column 3.

The Rumor Lesson Column 3 Selections

The team recorded notes related to their overall problem-solving approach. Often, the first teaching teams have no specific teacher questions to add. Many teams assign one observer to record actual questions posed by the teacher during the first lesson, giving the team a chance to discuss questioning at the post-lesson discussion, and providing material for enhancing Column 3 for future teachings.

Lesson Outline Problems/ Activities	Possible Student Responses	Notes for the Teacher
1. **The teacher poses the main rumor-spreading problem to the class. Students will work individually, then in groups, then share results. (15 minutes)** *The Rumor Problem: Tara, a high school junior, wants to ask Billy to the semiformal. Today, Tara tells her 3 best friends her plans. On the next day, her 3 friends each tell 3 of their friends. The rumor continues to spread in this manner. How fast is the rumor spreading? Model how the rumor is spreading over the first 5 days.*	Modeling for five days: Students will use a variety of methods • Graphing • Multiplying $3 \times 3 \times 3 \ldots$ on calculator. • Tree diagram • Drawings • Manipulatives Possible student errors: • Summing the numbers • Using a linear or quadratic model Day 9 Prediction: Students: • may ignore directions and continue multiplying by 3 • will be stuck and need a hint. • will use 3^9 to obtain the answer.	Encourage use of varied methods and materials. Ask students to recall TACKLE IT[4] problem-solving strategies. Allow students to learn/discover on their own— encourage some struggle before offering help. Encourage discussion of different methods within groups. Teacher: Circulate to locate different key approaches for presentation to class.
2. **After sharing various representations for first five days, students are asked to work in groups to predict Day 9 of the rumor, then discuss predictions and methods as a full class (15 minutes)** *The Rumor Problem, Part 2: How many new people will hear the rumor during the ninth day? You must predict for the ninth day without calculating the sixth, seventh, or eighth days.*		

Imagining yourself as an outside observer preparing for the Rumor Lesson, what *can* you learn about the team's pedagogical approach by reading their entries in Column 3?

Column 4: Evaluation of Learning

The primary purpose of the Evaluation column is to focus observers on the key outcomes and understandings the team is hoping to see, and to remind the teacher of things to watch for as the lesson progresses. Filling in Column 4, the team is asking:

---

4   *TACKLE IT* was a term used in a prior lesson to recall various problem-solving strategies.

What should observers focus their data collection on? What is evidence of understanding or learning at key points in the lesson? Mid-lesson formative assessment questions are often included.

The Rumor Lesson Column 4 Selections

Lesson Outline Problems/ Activities	Possible Student Responses	Notes for the Teacher	Evaluation
1. **The teacher poses the main rumor-spreading problem to the class. Students will work individually, then in groups, then share results. (15 minutes)** *The Rumor Problem: Tara, a high school junior, wants to ask Billy to the semiformal. Today, Tara tells her 3 best friends her plans. On the next day, her 3 friends each tell 3 of their friends. The rumor continues to spread in this manner. How fast is the rumor spreading? Model how the rumor is spreading over the first 5 days.*	Five-Day Modeling: Variety of methods • Graphing • Multiplying 3 × 3 × 3 . . . • Tree diagram • Drawings • Manipulatives • Possible errors • Summing the numbers • Using a linear or quadratic model Day 9 Prediction: Students will continue to multiply by 3 or get stuck and want a hint or use 3^9.	Encourage use of varied methods and materials. Ask students to recall TACKLE IT problem-solving strategies. Allow students to learn/ discover on their own— encourage some struggle before coming for help. Encourage discussion of different methods within groups. Teacher: Circulate to locate different key approaches for presentation to class.	What methods are students using? Are students correctly modeling the exponential growth? Is exponential notation being used? Did students use new methods to predict Day 9? What new methods are being used? How are students communicating their methods and solutions?
2. **After sharing various representations for first 5 days, students are asked to work with group to predict Day 9 of the rumor, then discuss predictions and methods as a full class. (15 minutes)** *The Rumor Problem Part 2: How many new people will hear the rumor during the 9th day? You must predict for the 9th day without calculating the 6th, 7th, 8th days.*			

▌ 3. Summary of Findings

After the post-lesson discussion, and after team members have had a chance to reflect on what they have learned, a final section on Team Learning, or Research Findings is usually added. This section of the research lesson plan is generally filled out at the end of the cycle as the team considers and summarizes what they have learned. It plays a critical role for the *research* and *dissemination* functions of the plan. Much more information about how the team analyzes *What Have We Learned?* and about *Sharing the Learning* are found in Chapters 12 and 13.

The Rumor Lesson Findings

The first teaching of the Rumor Lesson was an eye-opener for the team. As predicted, students did apply a variety of methods to model five days. However, the requirement to predict Day 9 without calculating Days 6, 7, and 8 sparked a whole array of unanticipated creative strategies like "multiply Day 5 by Day 4 to get Day 9." Each of these responses reflected an intuitive knowledge of the laws of exponents—in this case $a^b \times a^c = a^{(b+c)}$. The lesson also revised the team's thinking about what their students had learned in prior years. Students used *no* exponential notation or formal laws of exponents that they had learned in their eighth-grade class, yet did exhibit strong basic understandings of exponents. For the second teaching, the investigation of Day 9 became the central focus of the lesson because it provided an opportunity for students to think about the exponential pattern and the connection between repeated multiplication and exponentiation. One teacher's end-of-cycle reflections on the lesson summarize her learning from the lesson:

> While teaching this lesson our group of teachers made some very interesting discoveries. . . . When students were asked to find the number of people that learned the rumor on the ninth day, without using the number of people who learned the rumor on days six, seven, or eight, many started using properties of exponents although we can be almost certain that they did not know these properties of exponents. . . . It became clear to me that students could learn mathematics quite naturally when they are prompted to do so in a way that makes sense to them. Instead of giving the students the rules of exponents and showing them how they can be used to acquire the answer to the question, students were able to develop these properties based on their previous knowledge. I learned that students could be naturally mathematical when they are allowed to do so in a way that feels natural to them. I also learned that teachers are at an advantage when they are allowed to share their results from lessons with other teachers. In this way the lesson can be modified and developed with students' expected responses in mind.

When Is the Research Lesson Plan Ready for the First Teaching?

One challenge teams face when writing a research lesson plan that includes the components described in this chapter is a tendency to try to do too much when completing it before the first teaching and observation. Many new teams use up most of their meetings planning in great detail for the first teaching, not realizing how informative it will be to observe the lesson. Envisioning the first teaching as more of a trial run gives the team a much greater opportunity to refine the lesson plan based on real student responses and to reflect on team learning at the end of the cycle. Furthermore, in trying to think through the difficulties students might encounter in the lesson, some teams go overboard in fully "armoring" the lesson. If the team wants to learn how students are thinking and how they develop understanding, it might be more helpful to surface errors and misconceptions during the lesson than to provide students with a fully lit path that has no curves or bumps. The takeaway is that teams should consider the lesson plan a work in progress throughout the cycle—they will start filling it in during the initial lesson development, but will continue to fill in new ideas and wonderings and to refine the rationale and lesson design as they teach and observe the lesson and as they reflect back on the cycle to summarize their findings and learning.

Questions for Deepening Practice

The *lesson design* (as summarized in the written plan, and as taught to students) represents the team's ideas and hypotheses for effectively teaching the lesson content. In this way, the *written lesson plan* is a tool the team uses to teach the lesson they have planned together, to engage in research that results in their own professional learning, and to share that professional learning with colleagues and the field of education. As your team works to deepen their lesson study practice, it is important to consider how well the team is developing research lesson plans that serve these three functions.

The Plan as a Teaching Tool

- Does our lesson flow help the teacher know what to do without feeling like a line-by-line script?

- Are our anticipated student responses realistic and useful?

- Are there follow-up questions, or suggestions for how to have students share their thinking?

The Plan as a Research Tool

- Does the rationale include a theory of mathematics learning?

- Will the lesson reveal student thinking?

The Plan as a Dissemination Tool

- Does the written document reflect and communicate the team's ideas?

- Does the plan include a description of our rationale for the lesson design?

- Does the written document reflect the most important debates our team had?

CHAPTER 8 | Anticipating Student Responses

The main purpose of lesson study is improved instruction. To accomplish that purpose, we pay very close attention to children and children's activities. We find out what children understand, how children think about the concepts that they don't understand, and who children are—what they are thinking and feeling. The more you understand what children know and how they think, the more effectively you can plan research lessons. This approach is always important, even in daily lessons. (Sugiyama 2005)

For those who have done lesson study, even once, the phrase *anticipating student responses* conjures up the point in the cycle when the team has finished choosing the key problems for their lesson, and turns to imagining how students will answer the problems and to recording this information on the research lesson plan. The task of anticipating student responses formalizes the focus on student thinking that pervades the entire lesson study cycle. We have chosen to devote an entire chapter to this aspect of lesson development because teachers and coaches have told us how generative the process is and how much it helps them in envisioning the teacher's role in the lesson.

Anticipating student responses means predicting what students might do, say, and think during the lesson, and why. This process prepares the team to teach the lesson, and strengthens their picture of the mathematics in the lesson. Anticipating student responses is also an important part of the research process in lesson study. What the team anticipates from students reveals their theory of how students will think about the mathematics and what they believe the important mathematics in the lesson to be. The team records the anticipated responses on the lesson plan so that lesson observers will learn how the team expects students to participate in the lesson.

One example will give a sense of the variety of detailed responses teams generate. In Chapter 20, *How Lesson Study Changed Our Vision of Good Teaching*, the team details student responses for several lessons, along with how the work of anticipating them impacted their teaching. One of those lessons was about understanding area as more than a formula. The main lesson task provided an engaging scenario, the rescue of a beached and wounded whale, and asked students to find the area of the wound so it could be bandaged. Anticipating how students would respond, the team hypothesized:

> Students might: tell you that area is a formula, that is, use some formula that they had already learned, or perhaps adapt this formula to the present picture; [or] trace the wound on graph paper and count the blocks inside; [or] trace the wound and subtract the squares

outside the wound; [or] divide the wound into rectangles or squares; [or] fail to convert square feet to yards correctly; [or] discuss the similarities and differences in finding the area of different irregular shapes; [or] discover more efficient methods to determine area: dividing the shape into squares, circles of radius 1 dot, or quarter circles of radius 2 dots.

Some of these methods had arisen during the teachers' exploration of the problem during their topic research. Others came from their experience of how students have thought about measurement of irregular figures in the past. Generating these responses helped the team think about what kind of understandings students might bring to the lesson, and what new understandings students might build in the lesson.

In this chapter, we explore the different types of student responses teams anticipate, how teams generate them, and how the process of anticipating responses can impact teaching of the lesson.

▌ What Types of Student Responses Do Teams Consider?

Anticipated student responses include things students might do, say, think, or feel as they tackle the lesson activities and mathematics. These include:

1. Approaches, strategies, and solution methods that students may use in the lesson, including standard and alternate algorithms, naive and advanced approaches, various representations, and false paths students take when they have never encountered the material before.

   ▪ Examples: 1. The rules students may find for the pattern are . . . 2. Students may have learned a trick, shortcut, or alternate way of representing multi-digit multiplication that works well but does not look like the method the teacher is suggesting. Why it works may not be transparent.

2. Common mistakes, misconceptions, or stumbling blocks that could influence students' thinking about the mathematics of the lesson, including thinking errors that are common with students who have never seen the material before.

   ▪ Examples: 1. Thinking that area is the formula and failure to deal with questions of unit conversion, or to express the answer in units. 2. Confusion about the words opposite and adjacent in first learning about the trigonometric ratios sine and cosine.

3. Important mathematical ideas and connections students may generate during their problem solving and during their discussion of the mathematics.

   ▪ For the injured whale problem, students may propose various methods of approximating the answer. The same approximation method may yield different answers when wielded by different students. Can we determine which answer is more reliable? How would one improve the approximation? Is there an iterative process they might use?

4. Ways students may use manipulatives and tools.

    ▪ Examples: 1. Students may need a physical cube to hold and manipulate in a lesson on surface area. 2. Students may use the graphing calculator to produce polynomial or exponential curves but improperly set window and get a distorted impression of the shapes. 3. Algebra tiles will help students start problems and check their answers, but students may prefer to show their solutions in algebraic notation. 4. Providing colored markers for the explanatory posters may result in groups spending a lot of time producing a gaudy Technicolor display.

5. Students' likely nonmathematical responses to the lesson (i.e., what will be engaging or frustrating, how they might work with their peers, which responses can be related to the team's broad goal for students—such as willingness to try something difficult without giving up). If the students reach some sort of impasse, how will they occupy their time?

    ▪ Examples: 1. Students will enjoy helping their partners, but may be reluctant to correct errors. 2. Students may respond with boredom if they think they already studied this topic last year. 3. Gender dynamics in groups may affect how students collaborate, including how mathematical disputes are settled.

How Do Teams Generate Anticipated Responses?

Thinking about how students will respond has been one of our most important areas of professional growth. How do we think of them? Many ways. Doing the activity together generates a lot of our ideas because we all approach the activity differently and we share our interpretations and responses. We also think about past students we have had and how they may have responded. We always miss something important, and that's where the first teaching is helpful.

As this quote suggests, the process of anticipating student responses draws on multiple experiences and processes. Some of the most common and helpful ways to spark the process are described here.

Exploring and Understanding the Mathematics of the Lesson
One teacher comments:

If I go in to teach a lesson that we all tried to do together beforehand, the other teachers solved the problem in more ways than I did. So when I go into the classroom, I can understand more of the students' methods of trying to approach the problem . . . I think lesson study gives more respect to the students' different approaches and the way they think.

Most teams begin by solving the lesson tasks, and then pushing themselves to generate approaches beyond standard methods and beyond each teacher's first method of choice. Taking time to talk about why each of these methods works will help the team understand the mathematics better, and imagine other ways of approaching the problems. Looking for unconventional methods, or naive student approaches, might help to recall what students were doing in the prior unit, and ask, "How would I approach this problem, if I had just finished that last chapter, but had never seen anything like this problem before?" And lastly, it is good to keep in mind that this is an iterative process. Exploring the mathematics generates possible student responses, which in turn generates questions for the team to explore further, and so on.

Drawing on Teaching Experience

Doing problems together rarely generates the most naive or incorrect approaches, but teachers do have classroom experience with how students typically react when faced with particular mathematical ideas or problems. That experience is another rich source of ideas for how students may approach the problems. Doing a group brainstorm on "How have we seen our students do these problems? What do they usually get stuck on? How do they go wrong?" should provide a set of responses that the teachers' mathematical explorations did not. Most teams are made up of teachers of all experience levels. This brainstorming highlights the veterans' classroom knowledge, but novice or out-of-field teacher will be able to offer useful perspectives as well. The result can be a fruitful knowledge sharing that also generates mutual respect for the value of different perspectives.

Using the First Teaching of the Lesson to Focus on Student Responses

Occasionally, the teachers on a team do not know much about a particular topic or are new to anticipating student responses, and need to use the first teaching of the lesson to generate student responses. This strategy works, and can be an explicit research goal of the first teaching. Whether the team approached the first teaching with this in mind, their vision of what to expect will be enhanced with the teaching of the lesson and should be incorporated into the team's thinking about improving the lesson.

Gathering More Data About Your Students

If the team is not sure what will happen with a particular problem, it's OK to try it out with a few students early in the lesson design process. Teams also sometimes rely on preassessments, analysis of student written work, or student interviews to generate more knowledge about how their students think about the mathematics. These pre-lesson trials work best if they are conducted in classes that will not be the ultimate observation groups, and should never be used to prep a group of students for the real research lesson. That would defeat the purpose of the research lesson, which is to maximize what we learn about how students learn and think.

Consulting External Sources

Lesson study teams also find value in turning to external sources for information about student thinking on the topic. Locating relevant sources of information is one role that leaders can play. Some useful sources of information for anticipating student responses include:

- *Journal articles.* National Council of Teachers of Mathematics (NCTM) journals often contain articles written by teachers that describe student reactions to particular lessons.

- *Published research lessons.*[1] Building on the research of other teams is a great way to learn about teaching approaches and student thinking on the topic.

- *Textbook teacher editions.* These may include sample student work or information on alternate methods.

- *Local experts.* A teacher, district mathematics specialist, mathematician, or special education consultant can be invited to a team meeting or review a draft of the lesson plan.

- *Lesson study resources.* Toolkits with rich resources on a particular topic are a new kind of resource for teams. See Mills College Lesson Study Group (www .lessonresearch.net) for more information.

How Does Anticipating Student Responses Impact Teaching the Lesson?

While the act of generating a list of possible student responses is rewarding and useful on its own, by helping teachers unpack the mathematics and make links to a learning trajectory for students, the most important part of the anticipation of student responses may well be the team's analysis and consideration of *what to do with these responses in the lesson.* The team now asks: *What mathematical and pedagogical power will these student responses bring to our lesson?*

- The actual mathematical territory of the lesson becomes better defined to include what you introduce and what the students may introduce. This wider territory opens the door for more mathematical connections, and the teacher's

---

1   In Japan, published research lessons are readily available and are seen as an important resource for teacher learning and for lesson study teams. See Resources Appendix for links to sites or books containing research lessons.

knowledge of this territory allows them to lead classroom discussions that utilize these connections.

- Having in mind the potential range of responses allows the teacher to think in advance of ways to organize the sharing of responses into a coherent mathematical sequence, building to the main point of the lesson.

- Anticipating that a problem can be solved in a variety of ways may encourage a teacher to give students leeway to discover things on his or her own in the lesson, rather than feeling pressure to provide a method in advance.

- Having in mind a range of approaches to a problem, the teacher can recognize correct alternate methods that may enrich the mathematical reach of the lesson for all students. Similarly, the teacher may be more able to detect the potential in almost correct or seemingly off-the-wall responses and craft a response that recognizes these students' thinking and encourages them in a mathematically useful direction.

- As student responses are anticipated, the team can brainstorm about teacher questioning or pedagogical strategies and include them in the lesson plan.

- Armed with knowledge of potential responses to a key problem in the lesson, the teacher can move about the room during student problem solving and quickly note which methods are being pursued and by whom, and locate unexpected responses as well. This allows the teacher to be more purposeful in deciding whom to call on during the lesson.

The anticipated student responses that teachers are gathering, and the ways they are using them in the classroom lesson, are bits of mathematical knowledge for teaching. Anticipating these responses, and using the observation to develop a more realistic set, is one of the most successful ways to build your understanding of student thinking on the lesson topic.

PHASE 3 OVERVIEW | # Teach, Observe, and Discuss the Research Lesson

Phase 3 of the lesson study cycle is the time for your team to shift from discussing, studying, exploring, and drafting to teaching and observing the lesson in an actual classroom. The pace in Phase 3 can feel very different from the earlier research and lesson development phases of the work because now everything is geared around an observation in a real classroom, the focal point of the lesson study cycle. Observing and discussing the research lesson makes the lesson "real," rather than just a theory. Your team, and other guest educators if you choose to invite them, will see the team's ideas played out with real students, and determine the effectiveness of those ideas based on shared observation evidence. That shared observation evidence is used in a post-lesson discussion focused on what observers saw, what that evidence might mean, what the team can conclude, and what the team should study further. Research, in particular about mathematics and student thinking, is as present in Phase 3 as in any other part of the lesson study cycle. Through research and discussions earlier in the lesson study cycle, your team developed theories about how to develop student understanding of the topic. These theories will guide the team's decisions about the data observers should gather during the teaching of the lesson and will provide the framework for the team's discussions of those data following the observation.

Chapters in this section treat those things that are newest for many teachers in the United States and where opportunities exist for teams to strengthen and grow their lesson study practice over time: lesson observation focused on the use of evidence, the post-observation discussion of the lesson, and improvement of the lesson. The set of activities and questions that are central to the team's work in Phase 3 is described here, with notes about chapters for further reading.

Key Activities	Central Questions	Related *Leader's Guide* Chapters
Plan data collection and observation strategies	• *What rationale did we develop for the lesson during Phases 1 and 2?*	Chapter 9, *Observing the Research Lesson*
Collect concrete data of students' learning	• *What do we want to learn from the observation?*	
Discuss the data with your colleagues	• *What data will provide evidence of student thinking and understanding?*	Chapter 10, *Post-lesson Discussions*

Key Activities	Central Questions	Related *Leader's Guide* Chapters
Improve and revise lesson based on discussion of observation data	• *What evidence is there that lesson goals were met?* • *What lesson revisions do the data suggest?*	Chapter 11, *Improving the Research Lesson*

Leadership Focus in Phase 3

Phase 3 has a different pace and flow from the other phases of the lesson study cycle due to the observation of the research lesson in real classrooms. Leaders can help teachers navigate the main activities of the phase (i.e., teaching and observing, discussing the observation data, and improving the research lesson) and can monitor the inclusion of outside observers. In particular, potential areas of focus for leaders during this phase include:

- Promoting evidence-based discussion

- Keeping the focus on the students during the observation and post-lesson discussion

- Providing protocols for observation and discussion

- Participating as a learner

- Helping the team prepare if outside observers are invited

- Orienting knowledgeable others who participate in the observation and post-lesson discussion

- Being prepared to moderate the post-lesson discussion

CHAPTER 9 | **Observing the Research Lesson**

Introduction

After many meetings devoted to discussions of mathematics and pedagogy, the research lesson will be ready for students and the team will be eager to see it in action. How does the observation actually work? What does it look like? Usually taking place in as natural a setting as possible (i.e., taught by a team member to one of his or her classes), it is observed by the full team and often by additional invited colleagues. This coach's reflection offers an image of the lesson observation:

The Day of the Lesson

The team has spent many hours together, studying and planning their research lesson, and today is their first opportunity to teach it and see how students will respond. Walking into the classroom, one will see students settling into their seats and getting out their textbooks, but there may be a different energy in the room—a bit more excitement than usual. Curiosity about the extra five teachers in the room who have come to observe their class is mixed with eagerness to see what the "study lesson" will be like. For the team, there is a sense that the lesson is an experiment about to unfold that will answer many questions they have debated together.

As the class begins, students and visiting teachers will be introduced, and the focus turns quickly to the lesson mathematics. Some observers will be sitting with student groups at their tables, and throughout the class they will listen and write intently as students grapple with the problems. Their main goal is to see what the students know about the lesson topic from prior studies, what new understandings are emerging in the lesson, and what unusual or incorrect methods students were using on the problems. Capturing actual student comments and written work allows the team to base their post-lesson discussion on evidence. A few of the team members circulate about the room, comparing what is happening in different groups and watching for evidence related to the team's broad goals. One observer has been assigned to record the teacher's questions. This makes it possible for the team to eventually record some of them on the lesson plan as suggestions for others who might use the plan. Later in the lesson, student solution strategies are shared and recorded on large poster paper to help the observers see the solution methods clearly and provide additional data at the post-lesson discussion.

The lesson study observation is clearly an important and generative moment in the cycle—a moment the team has been working toward, and something that will provide information to frame the rest of this cycle and possibly motivate future cycles. The observation is:

- *The research experiment*—a time to test the team's ideas and proposed lesson design and to collect data about how those ideas work.

- The moment when the team's *discussions about pedagogy come to life* for all to see and evaluate. We begin to see whether (and how) the lesson achieved its goals.

- The place where *students' voices fully enter* the team's discussions and student thinking about mathematics is on display.

- The time when teachers are able to verify, revise, and *enrich their vision of the mathematical territory* of the lesson.

Given how new the experience of teaching and observing a research lesson is for all involved, most teams allot time to develop the new skills, protocols, and culture that support this part of the lesson study process. In this chapter, we offer practical information and examples of how teams have improved their observation skills, and we also try to put the observation into perspective with regard to its role in the lesson study process and its potential benefits to participants.

How Is Lesson Study Observation Different from Other Classroom Observations?

Teachers we worked with made comments like, "The only time my class is ever observed is for my annual teacher evaluation." Similarly, "In our school, we are encouraged to visit other classrooms, but it almost never happens because there isn't any time or coverage." There clearly are schools where these comments might represent the exception, but in our discussions with teachers doing lesson study, these have been the rule. In many schools, teachers can go through an entire career and rarely have the chance to see and discuss a lesson in another teacher's classroom. It is not surprising that teachers therefore welcome the observation feature of lesson study. However, the *teacher* of the research lesson might wonder, "Is the observation meant to evaluate me or to critique my teaching skills?" and want some reassurance that the purpose of the observation is research on teaching and learning rather than evaluation of the teacher.

Why Is Live Observation Really Important?

What's the point of having so many observers participate? Couldn't everyone teach the lesson in their own classrooms and then come together for a discussion? Wouldn't videotaping the lesson be easier? Actually, the reasons for everyone to observe a live lesson, together, are not only numerous but deeply important.

First and foremost, the common live observation is preferred because it best supports the team's opportunity to learn. In a typical professional development workshop, interesting features of instructional practice and content may be discussed, but it can be difficult to then apply this knowledge in the classroom. Lesson study helps with transference by enabling teachers to see the lesson in context with their students, talk together about what they all saw, and discuss the implications for the design of the lesson and for their teaching. You may not be the one who actually teaches the research lesson, but you are there to see exactly what happens with the students.

Second, the team's learning is based on evidence, similar to other professional research. The teaching of the lesson is the "experiment" with the teacher and students as participants. Observers in the classroom take the stance of researcher, not co-teacher, and gather data needed to justify the team's conclusions, and to help the team better understand their students' thinking. Practically speaking, having more observers provides more, and better, data—many eyes see a greater volume and variety of information. Observers are better set up to capture data than the teacher of the lesson or a video camera can:

- The teacher is too busy thinking about and doing the teaching to be able to focus on observing. She has to move about the room, target interventions, teach and guide the students, and she can rarely sit still and listen or watch. Also, students often carefully edit what they say to the teacher. The observer will be able to capture more of what the students are saying to their peers. Observers have the luxury of watching a few students through the entire class, tracking the development of their thinking and noting what is happening in the down-time of the lesson.

- Firsthand, live observations provide more detailed and accurate observations than video. The average school video setup of one camera filming the class captures only the teacher's voice and image. Even professional videographers are challenged to produce images and sound that allow us to know what students were saying, writing, and doing.

Preparing for the Observation

At the observation, the teacher, the team members who observe, any guest observers, and the students each have a unique job to do, different needs, and distinct viewpoints. The number of observers will vary greatly from lesson to lesson. The team members always observe; but sometimes a few additional colleagues participate; and sometimes the lesson is public and a larger group of teachers, perhaps from other schools, attend. Make some basic plans to respect the needs of all participants, to make sure that the team's goals stay central in the process, and to remember a common lesson study mantra: "Everyone comes to learn."

Teams usually allot one full meeting to preparations for the teaching and observing of the lesson. Even if the team has undecided questions about the lesson plan and tasks remaining, it is still worthwhile to shift attention to thinking about how to gain the most from the observation. At the preparation meeting there are two essential tasks—deciding your observation focus and supporting the teacher of the lesson—and a bucketful of last-minute things most teams try to accomplish, if possible, to make the observation go smoothly.

- *Choose your observation focus.* Thinking ahead about *what you most want to learn* from the observation will give you a greater chance of building understanding about your main research questions. Remind yourselves: What are our understanding goals? What were our biggest debates about the instructional approach? What are we most curious about? These will produce guiding questions for observers. For example:

  - What is the *nature of interaction or collaboration* in the student groups? What *mathematical questions* are students asking one another in the groups?

  - Are students *using* the term *variable*? What is their *understanding* of *variable*?

  - How do students *start* the problem-solving task?

  - What representations (graphic, symbolic, pictorial, etc.) did students use in solving the problem?

- *Help the teacher get ready.* Devote time to whatever the teacher needs from talking over his or her last-minute questions or concerns to finding materials needed in the lesson. The teacher should feel strong collegial support from the team—from trust in his ability to teach the research lesson to hands-on support with lesson prep. Sometimes, the final decision about who will teach isn't made very far in advance, and so there may even be a need to talk over specific parts of the lesson plan or anticipate scenarios that this teacher envisions for his class.

- This *final to-do list* gives a sense of the assortment of additional business the team might typically be thinking about as they prepare for observing the lesson.

  - ☐ Check with school front office to make sure substitutes were arranged.

  - ☐ Talk with the students. Inform them there will be observers and explain the purpose of the study lesson. (See next section on student preparation.)

  - ☐ Do a final check of the handouts: Do the math! Make sure any changes in wording of the problems and directions on the worksheet are all working.

  - ☐ Make a seating chart so observers can record data with student names.

Preparing the Students

Many teachers wonder how students will react to this new experience. Will they act normally? Will they speak up? Will their approaches to the mathematics be typical enough to provide us with useful information? Overall, students usually do react positively, even with some pride, and go about participating in a natural way. Clearly, every class, every student, is unique, but our experience in many research lessons allows us to share a few lessons we have learned.

Students need to be prepared for the day's events by their teacher, given a chance to ask questions, and be told what to expect: who is coming to their class, why, and what the students will be expected to do (i.e., participate in a regular math lesson). Letting students know that teachers won't be there as helpers, but will be taking a lot of notes, is a good idea. Students should also be told that by participating in the lesson they are making an important contribution to helping teachers understand how students think and learn about mathematics. Knowing that teachers are eager to learn how they think about the mathematics may even encourage students to be more open about their thinking. Students are often surprised and impressed that teachers wish to observe each others' classes to learn about effective teaching. Seeing teachers as learners, as collaborators, and as interested in how students learn, is a powerful and positive insight for many students.

Students may enter the lesson with a bit of nervousness and excitement, but quickly become engaged in the lesson and go about doing mathematics as usual. There may be some students who feel shy in front of observers, but there are at least as many who are excited about having an audience and speak up more than usual. What matters most is that observers can see students engaging in mathematics, learn about their prior understandings, methods they choose, and what students understand about the mathematics. To *help students get past their initial nervousness* we have seen teams plan a brief icebreaker—perhaps having the students and observers greet each other, or giving students a quick but engaging first problem, something tangible to do right away so students feel ready to tackle the major problems of the day. One fourth-grade class was given a choice of either introducing themselves to the observers, or singing a song. They were unanimous in choosing to sing and clap an energetic song about multiples for the assembled teachers, and clearly felt proud and energized by doing this and by seeing the appreciation of the observing teachers. A class of older students might not wish to sing, but the team can try to plan a launch for the lesson that is open to all students and gets them talking.

Most teams make a point of thanking the class at the end, often clapping or offering a small appropriate gift for their contribution. Many teachers meet briefly with their students after the lesson to talk about the class and get their reflections on the experience, using this as an opportunity to let them know that their contribution was appreciated. Some teams plan a snack for the class at a post-lesson meeting or party.

Fresh Eyes: Preparing the Guest Observers

Most teams invite at least one non-team member to observe with them, often a colleague with expertise in the lesson topic or team goals (e.g., a special education teacher

Pre-lesson Meeting with Outside Observers

Come to learn and contribute to the team's research!

Agenda

What is a research lesson?
   (Explain about lesson study: Not a model lesson, or evaluation lesson.)
What will observers do?
   (Provide protocols and observation questions.)
What are the team's goals?
   (Read the lesson plan and rationale.)
Discussion and Questions

Figure 9–1 Pre-lesson Meeting with Outside Observers

for a team with differentiated instruction as a focus). Teachers of the same grade or course, or teachers from a *different* grade, a coach or mathematics specialist, or the school principal also make good choices. The team has been working on the lesson together for some time and may be so immersed in one way of thinking that they can't see other alternatives. The outsider's role is to bring new insights and fresh viewpoints, but they also *come to learn*. It is especially important to meet with them ahead of time to explain the research lesson and their role. This could happen at a regular team meeting, or, at a special preobservation meeting that touches on topics similar to the sample agenda in Figure 9–1.

Conducting the Observation: What Do Observers Do? How Do They Collect Data?

Simply put, the observers watch students closely and take a lot of notes relevant to the specific observation questions developed by the team. It is common practice for the observers to walk about in the room to see what different groups of students are doing or to sit nearby one table and listen carefully to the dialogue. Often, observers record their data right on the research lesson plan. This allows them to be aware of exactly what the team anticipated, and to be able to easily discuss and report their observations in the post-lesson discussion.

The goal is to collect data that will inform the team about the effectiveness of the lesson in building student understanding and in meeting the team's broad goals. Observers jot down what students say, write, and do that reveals students' thinking and learning patterns. These detailed observations will become extremely important in making meaning from the data collected. For example, in an algebra lesson on slope of a line, the observer might record specific data about:

What words did the student use in describing the slope of the line—ratio? angle? rise-over-run? rate of change? Y = mx + b? How many different methods for finding the slope were seen in students' written work?

This detailed evidence sets the team up for later discussion, not only of the yes/no question "Did they understand?" but the mathematical question of "What and how did they understand?" In addition to data on student thinking, teams often find it useful to notice things that will help them improve the use of time or other more structural questions about the lesson. Sometimes this involves tracking how many students used a particular method or how long the class spends on each activity. One team, for example, collected detailed data about how quickly students moved through the assigned work, shown here.

The class started working on the three worksheets at 1:00. Worksheets 1 and 2 required constructing graphs for a pair of equations. Worksheet 3 asked questions to build understanding.
    After 8 minutes, a quick scan of the room showed:

        4 students were done worksheets 1 and 2 and were midway through 3
        8 students were working on worksheet 2
        6 students were still on worksheet 1

    After 15 minutes,

        9 students were completely finished and socializing
        6 students just finishing up worksheet 3
        3 students were still working on worksheet 2

    At 1:25, the teacher called students together to discuss the questions on worksheet 3.

Having this data made it clear that only three students really needed the full twenty-five minutes to complete the assigned worksheets and also showed the cause of the students' socializing. The team revised the lesson to reorder the worksheets, putting the most important worksheets (1 and 3) first. At the second teaching, the class was able to come together after fifteen minutes to process the important "understanding" questions, allowing even the slowest students to get to these questions in time, and keeping the faster students busy with the second page of graphs.

Some Strategies for Organizing Data Collection

- *Seeing and hearing.* Hoping to gain insight into students' mathematical thinking, teams strategize ways of getting close enough to hear students' normal (but perhaps quiet) voices and see what they have written in their regular (but perhaps small) handwriting on papers. Seating one observer at each student table works very well, and seems to be more natural for the students than having observers looking over their shoulders. Another strategy is simply to ask students to write with markers on poster paper instead of with pencils on a worksheet.

- *Focusing*. Some teams find it works best to assign each observer a task. The team may divide up the observation questions and team goals, allowing each observer to focus on just a few things. One person might keep track of the timing of events. Another might collect data on a specific pedagogical question of interest to the team such as, "What questions from the teacher created the most discussion among students about the mathematics?" Someone might try to sit nearby an "average" student, a student whose first language isn't English, or one with a particular learning style or special needs.

- *Tools*. Some teams use special tools to support data collection (e.g., an observation protocol) or enlist technological support (video, sound recording, photos of posters, etc.).

Improving the Effectiveness of Observations

How do teams develop greater skill and effectiveness over time in this critically important lesson study process? How can team leaders and coaches play a role in building the team's capacity to observe and collect meaningful data? A team can arrive at the first teaching with no strategy for how to observe, no focus questions, and even an incomplete lesson plan, and *still* gain a great deal of useful feedback and experience the lesson observation as powerful. Seeing a live lesson that you have planned together is eye-opening, and major successes and difficulties with the lesson will be obvious. Soon, however, these teams will realize that pre-observation planning and conscious attention to gathering rich data will pay off hugely in team learning. In particular, as the team develops greater skill in observing and prepares for the lesson more strategically, the quality of their data will improve. Better quality data better supports their post-lesson discussion, which in turn allows the team to draw more substantial conclusions about their goals.

The team, individual teachers, and leaders will, over time, find what works well and what matters most to them in the observation. However, most come to realize that getting the most out of their research lesson observations requires attention to the following:

- *Lesson design*. During the lesson planning process, incorporate features in the lesson design that challenge student thinking while also revealing student thinking, making student work visible, and making student discussions audible. These features will provide the team with something to observe.

- *Focus*. Before the research lesson, think carefully about what constitutes evidence of learning and build a plan (and questions) for the observation that will focus on this evidence and other evidence related to team goals.

- *Fresh eyes.* Always invite at least one person from outside the team to bring fresh insight. Everyone on the team has, to some extent, developed common expectations. The outsider will see things the team cannot.

Anticipating the Post-lesson Discussion

For many of us, observing research lessons was initially quite foreign territory, and it took time and many less than optimal observations to realize that the features just described are really important. We also gained experience over time with *anticipating the post-lesson discussions while we are observing* the lesson. For example, while observing we may see an unexpected approach being used by students. We can foresee the questions that this method might raise when reported at the post-lesson discussion and so we can try to capture additional data related to those questions. Similarly, if we recall an excellent post-lesson debate being sparked by asking, "Exactly where in the lesson did you see the most learning going on?" we are more likely to notice these critical moments in the lesson. All in all, the observation provides the necessary ingredients for the post-lesson discussion, and so thinking forward may be one of the critical skills to try to develop over time.

Conclusion

Preparation is key to maximizing everyone's learning from the lesson study observation and the subsequent post-lesson discussion. Of course, the team's attention to the design of the lesson is critical and has been the primary focus of their team meetings. Before teaching and observing the lesson, the team needs to attend to any last-minute arrangements and materials needed, choose who will teach the lesson if not already decided, and support the teacher of the lesson. In addition, the team needs to think about how they want to focus the observation, considering questions such as: *What are our observation questions? What data do we want observers to collect?* and *How do we want observers to collect that data?* The students of the class and any observers should also be prepared for the observation in advance. Careful attention to preparation often yields better data for the team to make sense of in their post-lesson discussion, improving the team's ability to draw useful conclusions from their research lesson.

CHAPTER 10 | Post-lesson Discussions

Debriefing [at a post-lesson discussion] remains the biggest, the best, the most important experience of lesson study. There is such power in sharing, such power in having eyes not involved in the progress of the class, so much to learn from each other.

The post-lesson discussion is a culminating moment in the lesson study process. There is much excitement about the observation of the lesson, and the team is buzzing about questions such as, *What was observed about students' thinking? Did the lesson advance student learning of the mathematics? How are we going to make sense of all that we observed?*

Purposes of the Post-lesson Discussion

The purpose of the post-lesson discussion is to share observations, to discuss what those observations mean, and to come to some consensus about what the team has learned. Post-lesson evidence-based discussions refine and challenge the team's ideas about how students think about and understand the mathematics of the research lesson. The discussion focuses on the goals for students that teachers set at the beginning of the cycle and the related observation questions determined by the team. Key questions that guide the post-lesson discussion are:

- What evidence is there that lesson goals were met?

- What insights and conclusions can we draw from our observation of students engaging with the research lesson?

- What have we learned about student thinking, mathematics, and the lesson?

- What new questions do we have?

- What improvements or revisions to the lesson do the data suggest?

In a post-lesson discussion, sometimes referred to as a *debriefing meeting*, teachers start by discussing the concrete data (e.g., student responses, prior knowledge, misconceptions) that they collected during the lesson. These might include comments such as "I saw students really struggling with starting the problem because many were hesitant and looking around the room," "One group of students had a really interesting way to approach the problem," "They thought about it this way. . . ." After sharing evidence of student thinking and learning observed in the lesson, the team moves to a broader discussion of students' mathematical thinking, and of how well the lesson met the team's goal. The post-lesson discussion also offers an opportunity to strengthen participants' understanding of the mathematics as they make sense of what they saw students doing. The discussion closes with an analysis of the implications for the design of the lesson, and connections to everyday teaching are considered. Teams are encouraged to refine their hypotheses about effective instruction, and identify new or additional questions that they may wish to explore in future cycles of lesson study.

▌ Structures for Post-lesson Discussions

The post-lesson discussion usually occurs shortly after the observation, either immediately following the observation, or after school on the day of the observation. It is helpful to include time for participants to review their notes, collect their thoughts, and reflect on their observations and questions before beginning the discussion.

The following people are involved in the post-lesson discussion: a moderator, the teacher who taught the lesson that was observed, the rest of the teachers on the team who developed the lesson, and sometimes additional invited guests. One of these guests might be asked to serve as a knowledgeable other[1] and provide a closing commentary at the post-lesson discussion.

The basic steps/components of the post-lesson discussion are:

Introduction—A brief overview that sets the tone for the discussion.

- The moderator gives a quick overview of the purpose, steps, and timing for the post-lesson discussion, including any relevant norms the team has set for the discussion.

- The moderator thanks the teacher who taught the lesson and the team members who participated in designing the lesson.

- The moderator encourages participants to provide evidence for their observations and comments, and reminds participants that the focus of the discussion is on learning. If there are artifacts of students' work (e.g., chart paper, student work sheets, reflection sheets) produced during the lesson, they should be available during the discussion.

- The moderator may remind participants of the team goals and observation questions.

Sharing of Observations—An initial sharing that allows everyone a chance to put an idea on the table to fuel subsequent discussion.

---

1   A *knowledgeable other* is a person outside of the team who guides and helps a team improve their lesson study work. Outside experts can provide subject-matter expertise such as mathematics content knowledge, knowledge about the lesson study process, or even group facilitation expertise. See Chapter 16, *Incorporating Expertise from Outside the Team* for more information on knowledgeable others.

- First, the teacher who taught the lesson shares a few observations. Then, other team members take turns sharing data from their observations along with any ideas and questions.

- Other observers (e.g., invited guests) share observations, especially any surprises, questions, and observations particularly relevant to the team's goals.

- Observations are recorded on chart paper or a blackboard.

Focused Discussion—An in-depth discussion of a few ideas.

- The team engages in an extended discussion about the ideas brought forward in the sharing.

- The moderator can choose one or two ideas, usually the ideas of most importance to the team, on which to focus.

- As particular issues are discussed, the team usually shares key choices they made in designing the lesson and the rationale behind their choices.

Closing—Summing up of key ideas and lessons learned.

- A final commentator, if there is one, shares comments on a particular aspect of the lesson, usually focused on the team's goals. The commentator may also share new knowledge or offer a new perspective on the lesson focus (e.g., going into further depth on the mathematics, discussing the use of technology in the lesson, or sharing research on student thinking about the topic).

- Ideas for revision are raised, discussed, and recorded.

This is not intended to be a rigid structure, but one with some flexibility. If the group is very small, for example, the sharing of data and in-depth discussion of ideas may flow naturally back and forth and a moderator might be unnecessary. Over time, teams usually end up settling on a structure that works for them. Still, most often it helps to get the data out on the table before engaging in conversation to make sense of the data.

If there is a large group of outside visitors or observers at the post-lesson discussion, the structure may vary somewhat. For example, one could use a fishbowl approach: rather than everyone sharing observations, a small group may be selected in advance to participate in the discussion while others observe. Another variation is to poll the large group for issues to discuss, rather than sharing everyone's observation data. Sometimes the observation and post-lesson discussion occur within a large professional event called a *lesson study open house*, which gives the team a chance to share their learning and receive feedback from a broader audience. In these large events, a variety of post-lesson discussion models are used, including panels or breakout groups. (See Chapter 17, *Public Lessons: The Lesson Study Open House* for more on this.)

Differences Between First and Second Post-lesson Discussion

Most teams teach, observe, and discuss their research lesson twice within one cycle of lesson study. The basic structure of the two post-lesson discussions is the same, but the content of each will depend on what actually transpires in the classroom and on the prior experience of the team in observing or discussing lessons. The emphasis in the first discussion is on sharing and reviewing observation data, with primary attention to understanding student thinking and how the lesson can be improved. If the first lesson has major problems (e.g., with timing, manipulatives, poorly worded questions, or misconceptions about students' prior knowledge), most of the first post-lesson discussion will be aimed at identifying areas in need of improvement. For several reasons, the second post-lesson discussion may be a more in-depth conversation than the first. It benefits from insights garnered in both teachings. The larger number of guest observers who often attend the second teaching bring fresh viewpoints and knowledge. And by the second teaching, most serious lesson flaws have been improved and discussion can focus more broadly on the mathematics, student thinking, and the team's pedagogical theories. Greater attention can be given to generalizing the team's learning about the effectiveness of the lesson, the mathematics, students' learning of the mathematics, and what the team has learned about teaching and themselves as a team.

An Effective Post-lesson Discussion

In an effective post-lesson discussion, everyone has clearly come to learn. Certainly, the team of teachers who developed the lesson hopes to learn from the lesson that they planned together. By sharing the lesson the team developed with others participating in the observation and post-lesson discussion, the team creates an opportunity for many others to learn from the lesson as well. In our work, we have noted a few features that contribute to an effective post-lesson discussion and have guided our work as coaches and facilitators working with lesson study teams. These features include:

- Designating a moderator or facilitator who assumes leadership for the discussion

- Creating a safe space for inquiry and analysis

- Using evidence to support conclusions

- Focusing the post-lesson discussion on mathematical learning

- Recording the discussion and lessons learned

Designating a Moderator or Facilitator Who Assumes Leadership for the Discussion

Post-lesson discussions are usually moderated by a teacher from the team, a coach, or another person who is familiar with the team's work and with the goals and processes of lesson study and the post-lesson discussion. A primary role of the moderator/facilitator is to work along with the lesson study team and any participants to ensure that the post-lesson discussion is effective and supports participants' learning. The moderator facilitates the group's learning by keeping the discussion focused and moving forward. If there are outside visitors or participants not familiar with the lesson study process, the moderator should comment on the purposes of the post-lesson discussion and explain the structure of the discussion.

Creating a Safe Space for Inquiry and Analysis

Experimentation is central to the process of lesson study and because instructional ideas and approaches are being tested in a public fashion, teachers can feel nervous or vulnerable when sharing their ideas and lessons. Within a post-lesson discussion, it is critical that the discussion focus on the ideas being shared rather than on critiquing the teacher who taught the lesson or the teachers who developed the lesson.

One of the goals for a moderator/facilitator in the post-lesson discussion is to encourage inquiry and analysis while ensuring a comfortable learning environment for all participants. He or she might remind everyone that the primary goal of the discussion is learning about mathematics, student thinking, and instructional improvement. The tone should be respectful, particularly of the teachers who have developed the lesson. However, sometimes the discussion can err on the side of being "too polite" where participants are reluctant to raise issues or questions that might be challenging or viewed as critical of the lesson or the teacher. This doesn't enable real discussion of the mathematics of the lesson or student thinking and learning. Moderators and participants need to build a sense of trust and community in the discussion, acknowledging the teachers' contributions and their risk taking while encouraging participants to share their observations, particularly surprising or unexpected observations, and creating a discussion that invites a lively exchange of ideas.

The following strategies can help a moderator support a productive discussion:

- offer guidelines for the discussion at the beginning (e.g., suggest that participants refer to the lesson as the *team's lesson* or *our lesson* rather than as an individual's lesson)

- prepare the team ahead of time for the kinds of questions they might be asked (e.g., why did you decide to design the lesson they way you did?)

- move the focus toward the effectiveness of the lesson design and student thinking

- model the tone you want to set when asking your own questions

- build trust and community among participants by directing questions and comments to the team rather than the individual teacher, using *we* when the team discusses the lesson, for example

Using Evidence to Support Conclusions

Lesson study is a research process, and the learning that results from the process is grounded in the collection and analysis of data on student thinking about the mathematics central to the lesson. By basing conclusions and insights on evidence, the ideas that are learned through lesson study are not just opinions or individual perspectives but ideas gained through the shared experience of observing and making sense of students' learning processes. One of the challenges for the moderator is that observers' comments can sometimes be quite broad or general (e.g., "I think the lesson went really well!") and he or she will need to probe for specific data that prompted the observation. If the lesson went very smoothly it would be easy to celebrate and call it a day. In this case, the team may not learn much unless they probe deeper: talk about what *exactly* *made* it go so well, think about what they have learned, what surprised them, what they saw students doing and heard them saying. The team can extend or deepen discussions by trying to go beyond global statements and making explicit the evidence that support their conclusions. What might this sound like? A statement like "The lesson went really well" could be extended as follows:

> The lesson went really well! Our goal was [insert goal here]. And I saw students doing [insert data here]. In addition, we were wondering [insert research question here]. And I think our lesson design helped us learn about this question because [insert data here].

The following kinds of questions, asked by team members or the moderator, might help everyone to be more explicit about the evidence and data that underlie their assertions. What did you see students doing? What did students do or say that showed they were [engaged, understood the main ideas, etc.]? What evidence do you see that students were working toward the team's goals? What did students struggle with, or what questions do you have about their understanding of the mathematical topic based on what happened during the lesson? What evidence did you see or hear of what the students learned from this research lesson?

Focusing the Post-lesson Discussion on Mathematical Learning

When the lesson is taught, quite a bit of observational data is collected. In order to provide focus when using those data, the key question to consider in the post-lesson discussion is: What did we learn about the mathematics and students' learning of the mathematics? It can be helpful at the beginning of a post-lesson discussion, when everyone is eager to share what they have seen, to take a few minutes for some individual reflection and writing on that key question. It is also important to keep the focus on students' mathematical learning throughout the discussion, and when summarizing at the end.

While there is still a reasonable amount of time remaining, the discussion should move from the sharing of individual pieces of evidence to *making sense* of what was observed. If this does not happen naturally, the moderator should facilitate a shift to analysis of the effectiveness of the lesson in supporting students' mathematical learning. To do this, the moderator can highlight important issues related to student thinking about mathematics or to the team's goals. Questions that coaches on our project found helpful in sparking and deepening discussion at a post-lesson discussion include:

- Did the lesson activities and flow of the lesson contribute to achieving the team's goal? What mathematics did students learn in this lesson or in a particular part of a lesson?

- What student mathematical thinking did the problems and materials help to elicit? What helped to keep students' focus on the mathematics? Were students' ideas encouraged in the lesson? Look at the lesson problems one at a time and discuss how students approached them.

- How did the classroom discussions and teacher's questions promote student understanding?

- What did students learn about mathematics? What evidence was there of this learning? *When* in the lesson did we observe the *most* student learning?

- Was the lesson content appropriate for the students' level of understanding? Did students apply their prior knowledge to understand the mathematics of the lesson?

The goal is to refine ideas about how students think about and understand the mathematics of the research lesson, which can later inform revision and improvement of the research lesson before teaching it again. As the moderator, try to keep the focus on the relationships between the mathematics, student thinking, and the team's goals by asking questions such as, "What was the mathematical focus of the lesson? What evidence did you see of student learning of the mathematics?" If you have an outside expert or knowledgeable other participating in the discussion, draw on their expertise and plan for or encourage them to contribute to the discussion with an eye on teachers' learning.

Recording the Discussion and Lessons Learned

It is very useful for the team to have a record of the conversation at the post-lesson discussion to support their continuing work together. Some ways that teams create this record include: assigning a note taker for the discussion, creating a visible record on chart paper of highlights from the discussion, collecting individual written reflections, or videotaping the discussion. In addition, shortly following the post-lesson discussion the team will often prepare a summary, highlighting lessons learned. These record-keeping efforts are invaluable for the team as they revise their lesson and consolidate their learning from the cycle.

We have experienced a phenomenon that we are sure other teams have noticed—that the wealth of data shared, the excitement of seeing the lesson live, and the power of discussing the lesson with colleagues can fully absorb our attention as participants. Recalling all the important points later may be difficult. Taking time at a later meeting to look at the notes or watch a video of the post-lesson discussion can be quite illuminating, and can be a powerful way to begin consolidating learning from the cycle and identifying future questions for investigation.

Conclusion

Improving the quality and learning from our post-lesson discussions take time and skill. Lesson study teams vary enormously in teaching experience, mathematics knowledge, and collaborative style among other things. A team's first post-lesson discussions may be limited due to inadequate observational data, or discussion focused on non-mathematical issues. Or, sometimes teams are overwhelmed by the power of observing students work and they have so much data to share that the discussion never stays focused on one thing very long. For any post-lesson discussion, the best advice is to strive for an open sharing of ideas and a focus on trying to learn from what was observed. The post-lesson discussions *are* opportunities to nurture a strong focus on mathematics, student thinking, and team goals but an equally important goal is for the team to have ownership over the discussions and be able to honestly discuss their reactions. It is the team's chance to make sense of what was noticed and make real improvements in the lesson. Over time, more advanced lesson study teams gain skill at collecting and sharing relevant observation data, analyzing this data, and identifying their learning about students, mathematics, and teaching from the lesson. The moderator/facilitator and knowledgeable others can play a critical role in helping teams develop these skills and building the teams' ownership in using the lesson study process to improve teaching and learning.

| # Improving the Research Lesson

There is a great sense of accomplishment and relief after a lesson study team completes its research, planning, writing, and first teaching[1] of the lesson. The lesson observation is often eye-opening, and reveals many things for the team to consider about its lesson. It provides evidence to support the team's hypotheses for teaching their mathematics topic or, alternatively, it provides evidence that prompts the team to question its theories. The next step in the process, improving the lesson in preparation for a second teaching, is a rich opportunity for team learning. Teams analyze data from the first teaching, consider what most needs improvement, and then redesign their lesson to better foster students' learning of the mathematics. The teams consider both their goals for students and the evidence of student thinking and understanding gathered at the lesson observation. Revisions also should take advantage of the team's research and analysis from the initial lesson development stages. In this chapter, we start with an example of how a team used evidence from their lesson observation to inform their revisions, address the different types of improvements that are often made to lessons, and then detail the revision process.

▌ Using Observation Data to Inform Revisions

Unanticipated responses or surprises observed in a research lesson often point to areas in the lesson that may benefit from consideration and improvement. Did students learn something or do something the team didn't expect? What unanticipated responses or misunderstandings were observed in students' work? If there were mathematical misunderstandings, the team should try to unpack those misunderstandings, working to figure out what student thinking led to them. Teams often find that their discussions about student misconceptions are some of their richest conversations. The team can add these misconceptions to their list of anticipated responses when they revise the lesson plan. They can also consider how the teacher might respond to them—does eliciting these student responses enrich the lesson or do they move the lesson away from the intended goals?

---

[1] This chapter is framed around the revisions after the first teaching, observation, and post-lesson discussion. Chapter 12, *What Have We Learned?* discusses the period of reflection and writing after the second teaching, where additional revision is considered.

> One team planned a lesson in which students were asked to find an algebraic rule for a number pattern. The team planned to give students a few minutes to start thinking about the problem, then ask them to share out their best guesses before returning to working out a full solution. This quick share-out was intended to help all students get started and to encourage multiple strategies. Of the ten student pairs, four or five shared out essentially the same method, but the other pairs were all pursuing distinctly *different* approaches. The teacher was pleased to see this variety, and the students returned to work. To the great surprise of the team, at the final sharing of solution methods, almost all of the ten pairs reported using one basic method. Had the students all concluded, after the first sharing, that the "most popular" method was superior? Or was it the easiest one to understand? The observers noted that the lack of variety had caused a real "deflation" of the mathematical discussions at the end of the lesson and the team realized that in this case their method of sharing different strategies early in the work period needed rethinking. A potential revision would be to eliminate the first share-out, but team discussion of goals and strategies for the share-out may help the team generate other productive ways to handle the situation.

Different Types of Improvements

Revisions after the first teaching are generally aimed at refining the lesson in the following ways:

- *Mathematical changes* such as improving the mathematical problems to make them more challenging or accessible to students

- *Pedagogical changes* such as altering the pacing of the lesson or the use of manipulatives

- *Changes to enhance the use of the lesson as a research tool* such as adding methods for making student thinking more visible or accessible to observers

Of course, in reality, these different types of improvements are highly interconnected. A change made to the lesson plan on any one of these fronts will affect the others. For example, making a change in the size of the student groups in a lesson can affect students' mathematical discussions. As the team makes revisions, try to keep all three types of changes in mind, being conscious of the interplay among specific revisions and the effect on students' mathematical understanding. The following descriptions provide different kinds of improvements along with examples to consider.

Mathematical Improvements

Mathematical revisions to the lesson are often prompted by questions such as, "What did students learn about the mathematics? What does the evidence show?" A team can examine data to see if their mathematical learning goal was met and to identify what parts of the lesson contributed to that learning. Adjustments to the lesson plan should be made to help ensure that students will be working toward those understanding goals in the future. Observation notes and student work are helpful tools for determining necessary revisions related to the student understanding goals. The team should consider data on individual and small-group interactions, and notes taken during whole-class discussions. A teacher-led student sharing at the end of the lesson often provides informal but valuable insight into what students learned during the lesson.

Mathematical revisions often involve changes to enable students to more directly engage with the mathematical ideas that are the focus of the lesson. Here's an example of how a team revised their lesson based on what they had learned about their goals for student understanding.

> In a research lesson focused on place value, the team had a theory that having students represent numbers in unfamiliar number systems such as base 4 or base 5 would highlight important number and place value concepts for students and ultimately improve their understanding of place value in base 10. The lesson was initially designed with different groups of students working to represent numbers in different bases by bundling straws. (For example, in base 5 the number 13 would be represented by 2 bundles of five straws and 3 single straws.) Some student groups worked with base 4, some with base 5, others with base 6. In the post-lesson discussion, the team reported that the students who worked with base 5 easily bundled their straws correctly while the base 4 and 6 groups were slower, making more errors, and failing to make connections as they worked. Also, all students had difficulty making generalizations in the full-class discussion, because the data from the group using one base, such as base 4, didn't make sense to the groups using a different base, such as base 6. The team revised their lesson to have all students work in base 5. Base 5 offered the contrast to base 10 that the team wanted, focused the students on the place value ideas (rather than computation), and created a common experience to facilitate class discussion.

Mathematical revisions can also involve making the mathematics problems in the lesson more challenging for students if there was a mismatch between the problem(s) and students' prior knowledge. One thing we've learned is that even very small changes in a problem or lesson can have a big impact on the mathematics involved and on student thinking elicited.

Pedagogical Improvements

Pedagogical revisions examine the instructional methods used in the lesson, materials such as worksheets and manipulatives, and the pacing of the lesson to make it more effective.

It can often be useful to analyze the impact of instructional methods used in the lesson design, in light of evidence of student learning collected at the observation. The team may then choose to revise teacher directions, the wording of questions posed to students, the use of the whiteboard or other materials, manipulative choices, and the level of differentiation for varied learners. For example, one team described a change they made to the specifics of their lesson plan as follows: "We found that with groups of four there were always one or two students doing most of the work and the others were just waiting for an answer. After the first run-through of the lesson we decided to change the groups from four people to two people." Here's another example:

> Another team developed a lesson on addition of fractions and they wanted to encourage students to use multiple solution methods. Rather than dictating the type of manipulative to be used, each table was given many types: fraction strips, cubes, graph paper, and so on. In the lesson, most students took the fraction strips or cubes, but didn't know how to use them to solve the problems. Students spent a great deal of time laying them out on the desks, lining them all up in rows, or playing with them, and as a result had neither the time, nor the method, for solving the addition problems. Disciplinary issues arose. Students didn't talk with their partners. In the revised lesson, all the manipulatives were made available on a table at the front of the room. Students solved the problems more successfully, and more quickly, in this lesson by talking to their partners, writing, and sketching their own representations.

Often, one of the first elements of the lesson that teachers think about revising is the timing and pacing of the lesson. In our experience, most lesson study teams initially plan too much for the amount of time set aside for the lesson, and so the team ends up streamlining the lesson plan before the second teaching. We advise teams to hold off on making any final decisions about how to streamline the lesson until they have considered the complete set of revisions they wish to make and prioritized the most important improvements.

Changes to Enhance the Use of the Lesson as a Research Tool

As team members consider what was learned about student understanding goals in the first teaching, they may find that not enough evidence was observable during the lesson to really know what students were thinking about or understanding. Students might have been mostly writing on worksheets that were difficult to see, or perhaps there was no discussion between students in small groups or in whole-class discussion. The team should revisit the lesson to determine improvements that might better reveal what students are thinking such as including a sharing out of solution methods, having students record their solution methods on poster board or large chart paper, and designing focused questions for a whole-class discussion to better understand students' thinking. These kinds of improvements overlap with the mathematical and pedagogical improvements described earlier, and frequently changes that enhance observers' access to students' thinking will help teams determine if their goals for student learning and understanding are met.

The Revision Process

In this section, we describe the revision process in more detail, offering suggestions for making it a purposeful and rewarding part of the lesson study cycle.

Note Ideas for Revision During the Post-lesson Discussion

Because it follows so closely on the heels of the first teaching of the lesson, the post-lesson discussion is a very natural place for teachers to think about changes to the lesson. This discussion provides a rich source of revision ideas; however, it is extremely important to *focus first on listening to the data reported* from all observers and *discussing what they mean* rather than diving immediately into what might have worked better. The team first needs to figure out what the data suggest *in order to* determine what improvements to the lesson are needed.

Practically speaking, the post-lesson discussion conversation may be the only time teams have for sharing observations and making sense of the data and planning revisions. In that case, the moderator needs to accommodate both conversations, keeping in mind the importance of basing the lesson improvements on the team's analysis of the observation data.

Gather Relevant Data and Resources

Team members will need to gather and bring relevant data that will help revise the lesson to the revision meeting. Data sources to have on hand will vary, but the following are often available and useful: notes from the post-lesson discussion, any chart paper used during the lesson, photos of the whiteboard or of students working, written student work, individual teacher reflections, and notes taken during the initial research phase of the cycle. Other preparations that may be useful include having the team review their observation notes, inviting an outside expert, and bringing their textbooks and other sources of mathematics problems.

Identify Which Changes Are Most Needed

During the team's discussion and analysis, participants will suggest many alterations to the lesson. It's very helpful to have one team member recording those suggestions. If everyone realizes that ideas are being recorded, the need to act on every suggestion is reduced. Lesson study teams sometimes approach revision by thinking: which changes will be an improvement? However, the first question should be, what most *needs* improving about the lesson? There are a lot of ideas that get floated like: I know of another problem that might be better, or let's use this other manipulative. It is important for the team to first identify the main things that need improvement, for instance student discussions and the students' engagement with the mathematics at the beginning of the lesson, before considering specific changes. As teams become more experienced, they will discover that once they have a clear picture of what the data show and ideas about the most needed changes, it's time to revise and record on an updated lesson plan.

Consider Different Types of Improvements

Several different types of improvements are often under consideration. Guided by discussions about the most needed areas of improvement, team members consider the revision ideas on the table. As discussed in the previous section, the kinds of changes under consideration might include: changes to improve the mathematics in the lesson, changes to improve the ways in which students engage with the mathematics, and changes that enable the lesson to be a more effective vehicle for testing the team's ideas. These different kinds of changes are interrelated and a particular change may offer more than one kind of improvement. In addition, the observation data likely contributed to the team's understanding of the teacher's role in the lesson. So there may also be improvements in teacher questioning or the teacher's role in the lesson that can be discussed. When team members offer their suggested changes, it is important to not only share the revision idea but also to explain why the change is likely to enhance students' mathematical understanding.

It is important for teams to keep in mind that this is a research process. Changes to the research lesson plan should be done with deliberate intention and with a continued focus on goals for student learning and the team's research questions. Ideally, the team should be able to observe students during the second teaching and be able to decide whether changes to the lesson impacted student learning.

Record Lesson Changes in the Written Lesson Plan

At this point, a lesson study team should make certain it has a copy of their first lesson saved. Usually, one person is designated to record the changes that the team decides on a new copy of the lesson plan (electronically or on paper). Try to document what changes the team makes *and* the reasons for the changes, in addition to making the actual changes that the team agrees on in the written lesson plan. It may also be helpful to revise student directions and teacher questions, add new anticipated student responses, and make changes in directions for teacher responses on the new plan. Finally, the team may need to revise observation tools to reflect the changes in the lesson. The original lesson plan, the revised plan, and the notes captured during the revision process including suggestions for changes and the reasons for selecting particular revisions will become important pieces of the team's document that captures their learning from the cycle.

What If the Team Sees No Reason to Revise the Lesson?

On some occasions, a team may feel that there's no need to revise their lesson because they learned a lot about students' mathematical thinking, or because they felt the lesson went well. Even in these cases, it may be worthwhile for teams to consider how they might improve the lesson design and take advantage of the opportunities for additional learning that the second teaching may offer them. A few scenarios follow for teams to consider.

The lesson provided ample data that the lesson design was fostering student understanding of the mathematics in the ways the team predicted. The team found many opportunities during the lesson to learn about students' thinking about the mathematics because the lesson's design allowed for student talk and writing. Based on this evidence, the team feels that the lesson design is "excellent" because they saw strong evidence of student learning. There is the temptation not to revise the lesson or even reteach it. However, even if no changes are made to the basic plan, it will be worthwhile to discuss how the needs of a second class might differ, and to observe student learning with a different group of students. Reflecting on the following questions may spark some additional improvements: Do we understand why was the lesson so effective? If not, how can we learn more about that in our second observation? What did the teacher do to enhance the prepared lesson plan? How might we capture those enhancements in the lesson plan? Note that the team's learning around such a lesson may be related to its place in the unit (e.g., the team might realize that the mathematics in the lesson is so critical that the lesson should come at the beginning of the unit).

The lesson was a "disaster"—it went way off the course set out by the team. Perhaps students were seriously off-task during the lesson. Or maybe students were tripped up by the wording of one question and unable to move beyond it. Perhaps the lesson drastically missed the mark in terms of students' prior knowledge. Perhaps the teacher had difficulty leading student discussion as intended. Even if the team wants to just start over, analysis of what happened during the teaching will be informative. Identify some small changes that could make a big difference and allow the team to revise and experiment further. Discuss possible changes—and then make decisions about what changes will have the greatest impact on helping to get closer to the team's goals.

The team doesn't have much data to discuss. Sometimes, a team reports that the lesson went fine and also doesn't have a lot of data to consider at the post-lesson discussion. When this occurs in a research lesson, it's usually a sign to ask more questions. Is this all the team has noticed? What specific evidence of student learning was observed? What data illuminates how those students were thinking or what they understood? Was the lesson challenging enough for students? One possibility is that student thinking was not readily accessible to observers during the lesson because students did not have sufficient opportunities to do mathematical thinking or because students' thinking was not visible to observers through the students' work. Another possibility is that the team didn't yet have a clear hypothesis underlying the lesson or didn't experiment with new ideas or methods in the lesson. The revision process offers the team a chance to create ways for students to share their thinking and a chance to rethink their observation process so that next time they have more data. Another possibility is that the mathematics problems need revising so that students are more challenged and can more fully engage in reasoning about the mathematics.

▌ Revision: A Simple Process with Rich Potential

Revising the lesson plan sounds like such a simple concept because it is all about improving the lesson plan based on an observed enactment of that lesson plan. However, the complexity comes in when the team works to unpack what will, in fact, make their lesson plan better and what *better* means.[2] The team will likely have dozens of possible changes to the lesson whirling through their heads after the rich experience of observing and discussing a teaching of the lesson. The team's job at this point in the cycle is to step back and think purposefully about whether particular changes take into account two important streams of information to which the team has had access so far:

- The team's goals for student understanding of mathematics content and broad goals for students. Particularly helpful to keep in mind are the team's research on the mathematical topic and their theories about how the lesson design will advance students' work toward those goals.

---

2   See Chapter 6, *What Makes a Good Research Lesson?* for more information on this subject.

- The evidence of student understanding, learning, and thinking (or lack thereof) collected during the observation of the lesson.

If your team can balance those two sources of information, and keep them at the forefront of their conversations during the revision process, then they will be well on their way to a rich and productive revision discussion, and to an improved lesson plan that will allow for continued learning for both teachers and students during the next teaching of the lesson.

PHASE 4 OVERVIEW | Reflect, Consolidate, and Share Learning

This final phase is about looking back and looking ahead. Your team will pull together the work they have been doing by identifying team findings, sharing these findings and other team learning with colleagues, and thinking about the implications for further research. Ideally, the team will reflect broadly on the cycle, considering the main areas of team work: their professional community, mathematics learning, research, student focus, and learning about lesson study. In each area, the team asks: What did we learn? How can we share that learning? What are our next steps?

This may be the most overlooked part of the cycle for new teams. Once the lesson is revised and retaught, it may be hard to shift gears from the active planning and testing work to reflection and recording. Or, the team may just run out of scheduled meeting time! Realistically, a new team also may not yet appreciate how useful their work could be to other teachers, or how writing a final report can be a learning *experience* for the team—not just final paperwork.

Topics in this phase include: Chapter 12, *What Have We Learned*? (articulating and consolidating findings and other learning), and Chapter 13, *Sharing the Learning* (ways of communicating with colleagues and the wider profession about the knowledge gained during this cycle). These are both areas where, over time, a team can grow dramatically in skill and in appreciation for the importance of this last phase of the cycle. The activities and questions that are central to the team's work in Phase 4 are described here, with notes about essays for further reading.

Key Activities	Central Questions	Related Leader's Guide Chapters
Finish analyzing observation data and summarize team findings about research questions and goals	• *What did we learn about mathematics, student thinking, and student learning during this cycle?* • *How did we revise the lesson and why?*	Chapter 12, *What Have We Learned?*
Reflect on the overall learning experience and benefits of the cycle	• *How has our thinking changed about teaching this topic or unit?*	
Revisit lesson rationale and lesson plan and incorporate changes	• *What new insights or knowledge have we gained that generalize to our other classes?*	
Share findings and learning with colleagues		Chapter 13, *Sharing the Learning*
Plan future research questions, cycles	• *What have we learned about lesson study?* • *What went well in our work this cycle? What do we want to change for the next cycle?*	Part III: *Building Sustainability and Connecting to the Wider Profession*
Revisit norms, and plan ways to improve team functioning in next cycle	• *What questions and ideas do we want to investigate in the next cycle?* • *How has our team grown as a learning community?*	Chapter 2, *Team Leadership and Group Norms*

▌ Leadership Focus in Phase 4

During Phase 4, the team will likely benefit from encouragement to keep their momentum going through the final tasks of the lesson study cycle and as they plan their future work together. In particular, leaders can help teachers navigate this phase by:

- Helping the team see their learning as mathematical knowledge for teaching

- Advocating for teachers as they plan for future lesson study work

- Helping the team get through the final writing tasks

- Encouraging the team to reflect and self-evaluate

- Encouraging the team to celebrate their work together

CHAPTER 12 | **What Have We Learned?**

Questioning and reflecting on what we have learned is not just an activity for the end of the lesson study cycle—it happens throughout the cycle at the individual and team level. But it is at the end of the cycle that the learning from all parts of the cycle is pulled together and used to draw conclusions. This is the critical time when analysis and interpretation of observation data occur and the team thinks about how their findings contribute to the profession. This is when the team asks, "What new knowledge have we gained and what are its implications for the classroom?" This is the time when the team goes beyond simple description of events and asks about their meaning. It is this probing of meaning that helps the team figure out how the lesson they planned really worked (or didn't)—not just whether it succeeded but how. It is this *how* that the teachers can pick up and carry with them to their other lessons or share with colleagues.

In sum, there is a critical part of the team's work that happens largely *after* the lesson has been taught. Building on data and analyses shared at the post-lesson discussion, this period of summative thinking identifies the team findings and directions for future research. In this chapter, we discuss the nature of findings and share examples of team learning, discuss the process teams use to consolidate their learning, and suggest some strategies for deepening practice during this important part of the cycle.

▌ The Nature of the Findings and Examples

The reflections and conversations about what has been learned touch on many aspects of the team's work. All in all the team is asking: What have we learned

- about mathematics and how we teach it?

- about student thinking and the trajectory of learning?

- about our goals?

- about lesson study?

During these reflective conversations, team members are trying to articulate what they have learned about their specific research questions and goals, but also to generalize and develop principles that will be useful in their own *and* other teachers' classrooms. Questions for further investigation are also generated as a natural part of this process.

We won't attempt to categorize the full scope of learning teachers can experience during a cycle of lesson study. Suffice it to say that teachers might walk away with new ideas and insights that range from a new understanding of a mathematical concept to a new appreciation of a colleague's expertise; from new images of students solving open-ended problems, to new ways to use a manipulative or organize a group discussion. Instead, the following examples are intended to illustrate the nature of learning teams experience around the main categories being discussed.

Understanding How to Support Students in Problem Solving

One team was aware that their middle school students had difficulty with problem solving, even though the teachers were sure that students knew a set of specific problem-solving strategies that were emphasized in all mathematics lessons. The team hypothesized that students were getting stuck as they solved the problems, and didn't know how to get unstuck. The team developed a research lesson to experiment with ways of helping students get unstuck. For example, the team provided students with clue cards. Each card contained a general question to ask when stuck, with the hope that this questioning would transfer to other times when they were stuck. At the lesson observation, much to the team's surprise, students used few of the clue cards or other aids provided. Faced with challenging problems, the students proceeded to solve without seeming to be stuck. But their answers were incorrect! Students weren't stuck. They were done! But they did not see their errors, nor proceed to try to verify or prove that their answer was correct. Similarly, one group of students worked independently and came up with different answers, but they were not sure how to figure out who was correct. During this lesson study cycle, the team learned that their students weren't stuck in the ways they had assumed, and proceeded to consider the next question: If students are not stuck, but done and wrong, what kind of teacher questioning would help move them off that spot?[1]

Recognizing and Interpreting Common Student Errors

Sometimes teachers recognize common mathematical thinking errors when they see their students making them in the research lesson. This shines a spotlight on the error, and is likely to be followed by team discussions about what it means and what to do about it. For example, one team noted in their lesson report under "What we learned":

> Use of equal sign 12 + 12 + 12 = 36 − 12 = 24 (Not OK!)
>
> What do we do to correct the use of these number sentences?"[2]

As with the last example, this team learning generated a new question: How do we help students use number sentences correctly? It could also generate a productive

---

1 Example summarized from case presented at AERA, 2004, by Brian Lord, Research Director, Lesson Study Communities in Secondary Mathematics.

2 Processing after lesson study, Chicago Lesson Study Group, Caterpillar Lesson, March 7, 2007. Lesson plan and notes posted at www.lessonstudygroup.net.

discussion about why students make this error. Is it a misunderstanding of the equal sign? Is it more related to students' language habits of speaking in run-on sentences?

Incremental Team Learning About Effective Use of New Pedagogies

As teachers explore the use of more open, problem-solving-based ways of teaching, learning what constitutes a good task or good teacher questioning is a long-term project. One teacher's comment, made at the start of the post-lesson discussion,[3] points to this realm of team learning:

I'm wondering if we led them too much?

Her team was experimenting with a more open form of instruction, but the lesson began with a set of guided exercises on a worksheet and a structured class discussion. It ended with an exploratory problem for which students completed a chart the team had provided. The teacher clearly hoped that observers participating in the post-lesson discussion would contribute to the team's understanding of the difficult question: How much direction is too much? A similar kind of learning statement was made by a team of teachers whose lesson is shown on the video, *How Many Seats*: "The students must do the work, not us!" (Lewis 2005).

Making Sense of the Mathematics and Student Thinking Behind Unexpected Responses

One team had been struggling to come up with a way to help their algebra students connect slope with the physical sense of steepness, and to find a way to represent this steepness numerically so that it made sense to them. After much discussion, the team decided to use the context of stairs because every student would have a physical feeling of the steepness of stairs. Students measured the stairs in their homes and made full-scale diagrams of them.

During the teaching of the lesson, as the class viewed all the diagrams, the teacher asked "Can you put them in order from least to most steep?" The team had anticipated that many of the students' stair measures would be very similar (because stairs are regulated by building codes) and had planned to play on students' uncertainty to motivate their next questions, "Is there a way to express the steepness with a number, to simplify comparisons? And how might we calculate such a number?" The first student spoke loudly and with great confidence, saying: "The Pythagorean theorem!" It took about ten minutes of skillful work by the surprised teacher to take this student's idea, find out what he was thinking, moderate the burst of discussion that followed, and eventually navigate back to the discussion of steepness measures.

---

3   Note: This lesson was taught at an open house with about twenty-five teachers in attendance.

As the team later reflected on this lesson, they discussed what this response could teach them about the student learning trajectory. They considered questions such as:

- Was it surprising that the student responded with something he knew about right triangles when he was viewing stairway diagrams?

- Is this a common way of thinking for students learning slope? Is this a meaningful connection worth exploring in class?

- How exactly do right triangle measurement and trigonometry relate to slope? What applications express incline using angle measure?

This student comment and the ensuing team discussions expanded their vision of the prior learning (or mathematical connections) students might bring to their study of slope.

Making Sense of Surprising Student Behavior

Lesson study leaders and coaches also reflect on their learning from research lessons. Here, one lesson study coach[4] relates a story about what he learned from a middle school research lesson, where he gathered data about a group of lively students.

> The research lesson included a mathematics problem for the students to solve called the handshake problem.[5] I observed a team of four students when they were supposed to be working on this problem together. Everyone worked separately for about five minutes, and then the first boy announced that he knew how to solve the problem. He started to explain, but his explanation was lengthy and halfway through his exposition one of the girls announced that she had a better method. She started her explanation, which was likewise lengthy, and halfway through it, the other girl interrupted her. Halfway through the second girl's explanation, the first boy interrupted to repeat his first explanation. . . . Round and round it went. . . . As a mathematician, I could recognize that all three students had valid methods, but no student could get more than 50 percent of his or her method on the table. Shortly before the time ran out, the first boy who had offered an explanation was so frustrated at the lack of attention that he was getting, that he jumped to his feet to improve his chances at dominating the discussion. Halfway through the exposition, one of the girls sprang to her feet for her interruption as well. At the end of the period all three of the mathematical protagonists were on their feet, and the fourth student was still seated but shaking his head in disbelief. That night the three students with purported solutions went home and wrote out their methods in full. They came back the next day, all with the same answer, and all of them surprised and perplexed that all of them were correct but had different methodologies.
>
> Relating my observations at the post-lesson discussion resulted in a good laugh for all, and a number of possible explanations were offered. Was it the result of the very strong

---

4   Joseph T. Leverich, Lesson Study Communities in Secondary Mathematics EDC Coach.
5   If you come into a room of twenty people, and everyone shakes everyone else's hand once, how many handshakes would that be altogether? Extend to solve for a group of any size (n).

personalities of the team members? Are students of this age very combative when they feel they are being treated unfairly by peers? I keep returning to this surprising observation, even years later, and new explanations have occurred to me. For example, maybe the students had not encountered many problems for which there were multiple methods of solution. Perhaps, since the three protagonists were convinced that their own methods were right, they could not even listen to other solutions. Or is it possible that the students thought that the explanations their classmates were giving involved so many steps that they could not possibly be correct?

One thing that I have learned is that the more bizarre a situation seems, the more reasons you will need to advance to explain it correctly and fully. In this case, the follow-up question, "What else could it mean?" or "What else could we learn from this?" might have provided some significant insights for the team. Other questions I would now suggest to teams that are trying to make sense of student thinking or surprising behavior include:

- Did this activity promote student mathematical learning? Did it promote student thinking? How?
- What other kinds of student learning did the wording or structure of the activity promote? How?
- A team with a long-term commitment to improving the effectiveness of "sharing of solutions" in their lessons might ask: What additional data do we have that will tell us what the students *did* gain from hearing each other's explanations or methods of solution? How did the sharing of solutions happen in other student groups? Was this more effective? Why?
- A team focused on promoting mathematical conversations might ask: "What did we learn about what students need in order to process one another's explanations?"

▌Reflecting and Generalizing—What Process Do Teams Use to Consolidate Their Learning?

Launching the Team's Final Analyses and Reflections

In the post-lesson discussion, the team is doing some immediate processing of what they observed.[6] They are *describing* their observations, including some that are quite interesting or puzzling, and noting important shortcomings and successes in the lesson. These lists will occupy a prominent place during the team's later meeting(s) that are focused on what the team has learned.

Launching the Individual Teacher's Final Analyses and Reflections

After the post-lesson discussions, it might be helpful for team members to do some individual thinking about what has been learned through the cycle. Questions such as the following can guide these reflections:

6   See Chapter 10, *Post-lesson Discussions* for more information.

- Of all the data shared at the post-lesson discussion, what was the most powerful "aha" for me as an individual? Why?

- What did our research lesson teach us about our team goals and about students' mathematical thinking?

- Was my vision of the mathematical content of the lesson or student learning trajectories changed through the cycle? How?

- What in our lesson study work has helped me learn the most?

The "What Have We Learned?" Meeting

In our work with teams we have found that it's most effective to set aside *dedicated meeting time* at the end of the cycle for reflection, analysis, and conversation to consolidate and record the team's learning.[7]

It is useful for the team to begin their end-of-cycle analysis of learning by sharing individual reflections. This will raise topics that teachers find compelling, and will naturally lead into discussion of broad questions such as, "What do we know that we didn't know before?" or "We saw that the lesson improved, but what made that happen?" It is likely these questions will generate a lot of productive discussion because individual team members will have different viewpoints or contradictory interpretations of the lesson observation data. Whatever hot topics arise, an experienced team will also be sure to focus some discussion time on each of the major areas of learning mentioned earlier: mathematics and how we teach it, student thinking and learning trajectory, team goals, and lesson study. These teams also consider what they have learned that they would use the next time they teach the topic, when they teach tomorrow's lesson, or in their teaching practice in general. Strategies for creating a productive analysis are similar to those for having a good post-lesson discussion, and include:

- *Asking for alternative explanations.* What might the data mean? What *else* could they mean?

- *Pushing for backup of conclusions that are suggested.* How do the data support that conclusion? How exactly do those data show that students understood the big concept?

- *Relating the discussion to team goals and hypotheses.* Were our assumptions about students' prior learning reflected in the data? Were our theories about how students might learn the concept born out?

---

7   Many of the Lesson Study Communities in Secondary Mathematics teams held this meeting at an informal out-of-school site, and followed the thoughtful discussion meeting with an end-of-cycle celebration.

- *Probing for why? and how?* Why did the students have difficulty with the problem? How exactly did the lesson help students understand the concept?

The goal of all of this probing and discussion is not to make a comprehensive list of findings on every topic, but to focus on identifying a few things that the team has learned. These findings are what the team will remember in the future, and will want to share with colleagues. Findings don't have to sound earthshaking to be important. Often, learning stems from in-depth discussion of a small detail of the lesson mathematics, a perplexing student response, or questions the team has about the teaching. In some cycles, the most important learning is about the team's lesson study process or about how lesson study is impacting their professional lives.

Expected Products of the Team Analysis

Many teams create a concrete final product that summarizes their work. This might be a presentation, open house, article, written report, or the lesson plan (with team summary of findings) for sharing with colleagues.[8] This expectation can add an incentive or create a focus for the final analysis. There may be many kinds of learning the team has experienced, and having an audience in mind may narrow the field to something manageable. At a minimum, the team should record findings about the team's main hypotheses and goals on their final research lesson plan.

▌ Deepening Practice

As important, rewarding, and interesting as this consolidation of learning is, it is surprising that some lesson study teams skip it altogether and many others have a difficult time with it. Asking the following questions may help your team get past some common difficulties and deepen your work in this final analysis and reflection period.

Are We Asking the Right Questions to Focus Our Reflection and Analysis?

Teachers' professional background may not have included this type of analysis of student classroom data. It takes time to develop a feel for what matters and for the kinds of questions that will spark discussions that elicit generalizable conclusions. As an example, the question "What have we learned about mathematics?" seems appropriate. But for many teachers, "learning mathematics" means encountering new college-level mathematics, not learning how students think about mathematics or learning more about the mathematics they teach. For those teachers, asking what was learned about the mathematics of

---

8   More information about the products of the team's analysis and how those products and the team's learning can be shared can be found in Chapter 13, *Sharing the Learning*.

the lesson topic or about student thinking on the topic might be more successful. We've given you many potential discussion starters and probing questions to get you started. Those lists may suggest possible questions that resonate with your team.

Were Our Goals and Research Questions Clear and Important to Us?

For some teams, lack of clear goals or lack of hypotheses about how students learn the mathematics, or lack of investment in those goals, makes it difficult to identify findings or conclusions. These are things that should have developed through team discussion during the beginning phases of goal setting, topic study, and lesson development. Still, it is not too late to bring these up when the team is making preparations for the observation, at the post-lesson discussion, or at the final "what have we learned" meeting. Asking what the goals and hypotheses were and what evidence of success would look like are always helpful in focusing the conversation.

Have We Looked at the Observation Data Through Multiple Lenses?

In the final analysis, your team is trying to make sense of what you have observed. To get the most out of this process, you might look at interesting or unusual individual responses to try to figure out what they reveal. Which observations were surprising, unusual, or interesting and how can we explain those data? Why did the students make that mistake? But looking at *the most common responses* is also important and gives the team different information. These may indicate what students have learned in their prior studies and suggest a typical learning trajectory for the topic or a common misunderstanding. In a similar vein, your team might want to seek new viewpoints from outside the team to help you view the data through a different lens and deepen your learning.

Are We Viewing Our Work as a Continuum of Learning and Incremental Improvement?

Many of the goals teams set address complex problems or seek change in fundamental teaching practices. Each cycle contributes some new ideas, and suggests new questions for further study. The goal isn't to identify a final answer or solution, but to make sure that you have mined the current experiment thoroughly, and realistically assessed what it contributes to your practice and other teachers' practice around the questions the team is studying.

Are Our Expectations Realistic?

Finally, always remember that lesson study is a complex process. Conducting research is one of many interconnected things your team is doing. You are also learning about the lesson study process and pursuing your professional learning in mathematics and pedagogy. Your team has set multiple goals for students (some mathematical, some developmental) that you wish to learn about, and your lesson study may be connected to a schoolwide initiative. Important learning from the cycle may occur in all of these areas. But the team does not have to approach this final analysis and reflection as a do it all project, nor as a one-time opportunity.

CHAPTER 13

Sharing the Learning

Teachers in different classrooms in the same school, or in different schools or districts, often struggle with very similar challenges. If a lesson study team has investigated how to help their students gain confidence in problem solving or found ways to help their students get a strong understanding of a key concept in algebra, it only makes sense for that team to share their findings with the many other teachers who face similar issues. Furthermore, the teachers on the team benefit, because, in sharing, they articulate clearly what they have learned, and open their work up for input and debate with a wider community of educators. Chapter 12 described the end-of-cycle reflection that every team does in order to articulate what they have learned. In this chapter, we consider how the team shares their learning with a wider audience in a way that will have the most impact on the team of teachers and on the wider network of educators and schools.

What Is Shared?

The Team's Research Lesson Plans

The research lesson plan is a natural written product of the lesson study process that can be shared with other educators. For example, one lesson study team participating in the Lesson Study Communities in Secondary Mathematics (LSCSM) project posted each new research lesson that they developed on the mathematics department website for their school. Other teachers from the school then adapted these lessons for use in their own classrooms. While a research lesson plan can be shared as a teaching tool, as this team did, it can also serve an important role in supporting the research of other teams. For example, if your team is thinking about how to teach about proportions, you can begin by seeking research lesson reports about proportions from other lesson study teams. Building on the research of these other teams provides a powerful way to gradually build and share new knowledge of teaching with the wider profession.

While the lesson plan is the most obvious product to be shared, teams should be cautioned to think about its limitations in regards to comprehensively sharing what the team has learned.

> The primary purpose of lesson study . . . is not to compile a library full of good lesson plans. This is useful, but it is actually a byproduct of the main goal of lesson study, which is to better understand student thinking and learning. A parallel goal is to foster teachers

as lifelong learners who continually seek to improve their teaching in the classroom. A library of good lesson plans may not improve classroom teaching if teachers are not educated as deep thinkers who can understand the rationale and objectives of those lessons. (Wang-Iverson and Yoshida 2005, 11)

When a team considers the lesson plan the sole product of their work, other educators may not be aware of, or have access to, the deep learning that the team has engaged in during the lesson study cycle. It may be necessary for other teachers to see the lesson plan in order to understand what the team did, but the plan by itself may not be sufficient to convey the team's learning. Having the opportunity to see the lesson live, and/or to discuss the lesson rationale and findings with the team is even more powerful.[1]

The Team's Learning About Mathematics and Student Thinking

If the goal of lesson study is to understand student thinking and learning better, teams should try to share specific insights they have gained, and the things students did, talked about, and misconstrued that provoked these insights. The learning that a lesson study team might share with a wider audience includes reasoning and ideas about how and why they developed the research lesson in the way they did, what a lesson with particular mathematical and broad goals for students might look like, student thinking and methods related to that lesson, and findings about the team's research questions and goals. Specifically, some components of the team's work that they might want to consider sharing include:

- New understandings that team members have gained—about mathematics, about lesson study, about the team's goals, about teaching, and about their students.

- The team's theories about the most effective way to teach the topic, and the reasons for the major choices in the lesson design. How did these choices stem from the team's research about the teaching and learning of the topic?

- Team and individual reflections about what has been learned about the team's research questions and goals.

- Description of the revisions made in the lesson and why those revisions were made.

- The learning trajectory for the lesson topic that the team has researched and refined.

---

1   See Chapter 17, *Public Lessons: The Lesson Study Open House* for more information on this model for sharing learning.

The Team's Experience with Lesson Study

While the primary work of the team is the cycle of inquiry about improving instruction, most U.S. lesson study teams are relatively new to the process, and may be the only team in their school. Many teams will take time to share their lesson study experience with interested educators to help develop a wider local network of lesson study teams, or simply to share with colleagues about their work. Sharing of learning about the lesson study process can happen as one part of the broader sharing of learning about the research lesson, as described previously, or it can occur as a special targeted effort. For example, to share what they know about lesson study, some teams mentor or coach new teams or host open houses to which they invite interested educators who want to learn about lesson study.

Who Might Be Interested in the Team's Learning?

As your team thinks about who might be interested in hearing about your learning, remember that lesson study is based on the philosophy that "everyone comes to learn"—it is not a spectator sport. Seeking an *audience* to hear about your learning also means seeking people who have something to contribute to your team. Many different people have reasons to learn from your team's work, and they have new ideas and expertise to contribute, or research of their own that can build on yours.

- *A few colleagues*. The team may begin by simply sharing the lesson plan and accompanying rationale with colleagues in their school who teach the lesson topic.

- *The department*. If the lesson relates to a departmentwide focus or initiative like incorporating new technology, or teaching through problem solving, the full math department might engage actively in discussing and using team findings.

- *The school*. If the team's work relates to a schoolwide initiative, or if the team wants to introduce lesson study to the other teachers in their school, presenting at a faculty meeting may be a productive setting for sharing.

- *The district and neighboring districts*. Often nearby schools share common teaching challenges (e.g., working with high percentages of English language learners, high failure rates in Algebra 1 classes, or increasing access to advanced placement courses) so the team may wish to connect with teachers or lesson study teams from neighboring schools.

- *The wider profession*. Eventually, the team may wish to begin looking at broader audiences, making use of educational or lesson study listservs or regional or national conferences, journals, or newsletters. If they have chosen their topic because their students have difficulty understanding it, the audience interested in their findings is very large. How many elementary school teachers, for example, would like to hear your findings on teaching fractions or middle school teachers want to learn how to better teach about percents and proportions?

How Do Teams Integrate Sharing into Their Work?

Starting Small, Making It Doable

Even if the team isn't sure how they want to share what they have learned, the key is to be sure to share something. Once the team has started opening the doors for their learning to stream out to others, and for others' ideas to flow back to them, sharing will become a natural part of the team's work. At the beginning of the cycle, the team should discuss who might be interested in their research lesson, and brainstorm a few ways to share something of their process, thinking, and findings. Of course, the team should always try to do as good a job as they can on writing up the research lesson plan,[2] including at least one paragraph on what the team was trying to do with the lesson (i.e., the team's approach and why that approach made sense mathematically, pedagogically, and according to the team's goals) and at least one paragraph on findings. The Resources Appendix includes some sources of research lessons the team might want to study, but every team can also simply write or present in the form that feels most natural to them—just trying to articulate what their goals were, what they have learned, and what they think other educators may find enlightening and useful. As more and more teams do this, a format or journal for reporting this kind of practitioner research will probably emerge.

Tracking Learning and Planning for Sharing Throughout the Lesson Study Cycle

To make the sharing more doable, it is important to begin thinking about sharing early in the cycle. Who will we invite to come to one of our team meetings or to the teaching of the lesson? Have we set aside a meeting at the end of the cycle for the analysis and sharing of our findings? How are we planning to keep records of ideas as the team develops them, so that we will have access to that thinking and rationale at the end of the cycle? The more the team puts into reflecting on what they are learning *during* the lesson study cycle, and making some notes as they go along, the easier it will be to summarize and share their learning at the end, and the more informal sharing they will do along the way.

---

2   See Chapter 6, *What Makes a Good Research Lesson?* and Chapter 7, *Developing the Lesson Plan* for more information.

Methods of Sharing

The following are several strategies that teams use to share their research lesson plan, their learning about mathematics and the development of student understanding, and their experience with lesson study.

Put It in Writing

Written products that describe the work and learning of the team can be distributed through websites, professional journals, reports that reside at the school, and targeted distribution to colleagues. For example, after completing four research lessons on one research theme, one team compiled the four lesson plans into a booklet with team reflections about each lesson and what was learned. The booklet was distributed to teachers from their school and other districts who attended a Lesson Study Open House hosted by the team.

Present and Discuss

The team, or members of the team, can share about their work and learning with faculty in the school (e.g., through faculty or department meetings), with members of the administration, with wider audiences at education conferences, or at another lesson study group's meeting. The following are some examples of how real lesson study teams have shared their learning by presenting about it.

- *Sharing mathematics learning with the rest of the team's mathematics department.* A small team within a large department showed and explained their research lesson to the rest of their department colleagues at a department meeting. Everyone first explored the main mathematics problem of the lesson together, and then discussed the research lesson.

- *Sharing lesson study experience through presentations.* A cross-departmental lesson study team at an agricultural high school explained what lesson study is and then presented about their first few lessons at an all-school faculty meeting—launching a wider lesson study program in their school on the theme of more closely connecting the content of the academic and vocational classes. In addition, the team introduced their work to the State Department of Education, and at numerous vocational education conferences, sharing how their school had used lesson study to connect across departments and improve the mathematics instruction for their students.

- *Sharing across teams or across schools.* Teams participating in a large lesson study initiative often have opportunities to share about their research lessons with other project teams. For LSCSM teams, a highlight of project workshops was the time when two or three teams would sit together, tell about their lessons, share about a challenge or success they were facing with the lesson, and receive feedback from the other teams. Cross-team sharing can also happen within a school or district. Teams could observe each other's lessons or come together at the end of each cycle to discuss what they had learned about pedagogy, mathematics, and student learning in the new curriculum.

Host Open Houses or Public Lessons

All teams (even new ones) should consider the idea of hosting a public lesson.[3] It can be very powerful for all concerned. Remarking on this, one teacher summed up what many have felt.

> *I just wanted to mention that after learning how to teach by being locked alone with the students in a classroom for many years, nobody ever stepping into that room to see what was going on, I was really touched . . . it was really profound for me to be sitting there and watch the teachers observing [the lesson]. . . . I really see this as an opportunity, taking teaching out of the closet, a tremendous change in teaching—giving it a professional dignity that it hasn't had.*

Public lessons not only share the team's findings, but also provide the team with a way to extend their own learning through the additional viewpoints and knowledge the guests offer. The sharing of knowledge and of learning flows in all directions. One teacher whose team had just hosted its first public lesson, says:

> *It isn't as scary as I thought it was going to be. . . . You do all the work of preparing the lesson anyway, and it isn't just for us. It's worth it because you get feedback. It's like proofreading your own paper. . . . You don't see your own mistakes. You don't get that other opinion . . . I would suggest to do it just to have different schools and different feedback that will help you as a teacher.*

Many of the middle and high school teams participating in the Lesson Study Communities in Secondary Mathematics project held an open house during one of their lesson study cycles. To host an open house, a team invites from ten to twenty-five educators to attend a teaching/observation of their research lesson and to participate in a post-lesson colloquium about the lesson. Visitors have time to explore the mathematics of the lesson before observing students engaged in the lesson, and in many cases, an outside commentator leads a session about the mathematics of the lesson after the observation and post-lesson discussion are completed. More information about hosting open houses can be found in Chapter 17, *Public Lessons: The Lesson Study Open House.*

---

3   For more information, see Chapter 17, *Public Lessons: The Lesson Study Open House.*

Helping More Teachers Engage in Lesson Study

Introducing other educators to lesson study is one way for a team to share what they have learned. This can be done many ways:

- Invite teachers to attend an open house with you.

- Ask new teachers to join your team, to participate in a meeting or two to learn about the process, or to observe a research lesson.

- Invite student teachers or specialists (such as special education or ELL teachers or aides) to join your lesson study team.

In all schools, there will be teachers joining or leaving a team through changes in their teaching assignments. As these shifts in team membership occur, knowledge of lesson study spreads among participants throughout the school, from veteran to new teachers and from experienced to novice lesson study practitioners. Changes in team membership can be due to this natural ebb and flow of the teacher population in a school, or it could involve purposeful switching of membership on lesson study teams in the school in order to spread ideas.

Sometimes experienced teams actively mentor novice teams in their school. For example, a team of high school mathematics teachers who had participated together on a lesson study team for several cycles decided to help teachers in other departments start lesson study. Each of the mathematics teachers served as a coach for a team from one other department in their high school as they worked to build lesson study into a schoolwide initiative.[4]

▋ Final Thoughts

To achieve strong lesson study practice, teams must make their lesson study practice open so that others can learn from their experiences and so that they can learn from others. Sharing the team's goals, ideas, lesson plan, reflections, findings, and learning with other teachers is essential because it opens these ideas to debate, to testing, and to further development. Sharing the learning takes one lesson in one classroom and makes it a contribution to the professional literature on teaching and learning mathematics—that is, it adds to the professional knowledge base on teaching. As your team goes forward, consider how you can both contribute to this professional knowledge base and benefit from it by opening your lesson study practice and learning to share with colleagues and educators.

---

4   For more information, see Chapter 24, *Expanding Lesson Study Practice at Our High School.*

PART III

Building Sustainability and Connecting to the Wider Profession

CHAPTER 14 | Planning for Sustainability: Beyond One Cycle, Beyond One Team

Lesson study is an innovation for U.S. teachers and schools. As such, it requires the same work to achieve sustainability that is needed for all education reforms or innovations: building understanding of the innovation, developing necessary skills and leadership, finding resources, avoiding replacement by an even newer idea, moving beyond the first eager volunteers, and so on. Any team that wants to go beyond one cycle, to deepen their practice over time, or to share lesson study with colleagues in their district will have to think about these issues. Overcoming challenges to sustainability of any innovation, including lesson study, takes time, persistence, and motivation. The stories of how some lesson study teams have worked to grow and sustain their lesson study practice are included in Part IV.

Why Does Sustainability Matter for Lesson Study?

Lesson study has features that make sustainability particularly important—some might even say that for lesson study, sustaining the work over time is essential. Why does it matter so much?

First, it does take time for teams to gain understanding of the lesson study process and the kinds of impact they can achieve over many cycles. A new team might not realize the kinds of benefits that can accrue, or that ongoing work will help the team achieve them.

Second, the learning people can experience through lesson study involves major changes in the teachers' thinking and in the school's professional culture. These kinds of changes do not happen overnight. The team might develop a new vision of what good teaching is or of what constitutes a good lesson. Teachers may change the kinds of questions they think about before teaching a lesson. A new focus on student thinking may emerge, or a shift to seeing the students as collaborators in the learning process rather than just recipients. Teachers may come to place a higher value on professional collaboration and research when tackling difficult issues. All of this takes time. In fact, lesson study is about lifelong learning, about integrating ongoing research into professional practice. Hence, thinking about ways to establish and sustain lesson study in this spirit are major priorities for teams and leaders.

Finally, sustainability of lesson study practice matters because it creates the opportunity for the advancement of professional knowledge. As lesson study work builds over time, there is an accumulation of knowledge, at the team level and professional level. Teams link their work from one cycle to the next. The school/district community accumulates knowledge about lesson study practice, about the teaching and learning of mathematics, and about mathematics content. Teams write and share their findings with the larger profession.

What Does It Take to Make Lesson Study Sustainable?

Important features of lesson study practice that contribute to sustainability include:

- *Within individual teachers, coaches, administrators:* Commitment to collaborating with colleagues to study how students learn

- *Within the team:* Strong lesson study practice and teacher leadership

- *Within the school:* Connection to local needs and goals, mathematical and administrative leadership support

- *Within the district:* Commitment to ongoing support for lesson study, connections among teams across the district

- *Beyond the district:* Connections to outside expertise, opportunities to share work with the wider professional community

In thinking about what this might actually look like for any one team or school setting, though, we should be open-minded and remember that each team or school has to start where it is, and slowly develop the kinds of structures and commitments that foster sustainability. Also, what sustainability means will vary greatly from setting to setting. We are not suggesting that every district doing lesson study must have lesson study teams continuing indefinitely districtwide, in all subject areas and all levels, with administrative support at all levels, and focusing on single goals for an extended time. Many less comprehensive, less long-lived efforts can be sustained for periods of time that achieve important impacts.

- One small group of teachers may continue to work together over many years, with just enough support to do their work well, making important contributions within their school or content area.

- A district may have a long-term commitment to including lesson study as an element in major initiatives, but may not actively support school-based teams outside the initiatives.

- A school may have an ongoing lesson study program within one academic department, but not put a priority on expanding to other areas.

- A university, or a group of teachers from several districts, may have a cross-district lesson study network that sponsors lesson study events, public lessons, and ongoing cross-district teams.

In other words, in thinking about developing sustainability, one shouldn't think only about producing district-level, systemwide programs. Think instead about *what you want to sustain and why* and then consider how outreach to or support from these different arenas (individual, team, school, district, beyond) will help you deepen and sustain that work. In all cases, to fully realize the team's learning and growth, we encourage making some connections beyond the team.

To support teachers and leaders in these efforts, chapters in Parts I and II of this book focused largely on deepening the lesson study work of the team. In Part III, we share what we have learned about building connections and support in the wider school and professional communities. These topics include:

- Building Partnerships with School Administrators (Chapter 15)

- Incorporating Expertise from Outside the Team (Chapter 16)

- Public Lessons: The Lesson Study Open House (Chapter 17)

These chapters offer suggestions for developing sustainable lesson study work within the school or district community and for situating the team's work within the larger professional community—through opportunities for wider participation, access to expertise, and training for new members. In addition, the later chapters in Part IV offer narratives that show a variety of ways teams and leaders have shared their work with the wider community and developed a wider lesson study program in their schools.

CHAPTER 15 | **Building Partnerships with School Administrators**

Administrators are among the essential partners in every school-based lesson study program.

Although the work of experienced lesson study teams is teacher initiated and teacher directed, administrators *can* and *should* still play active roles in supporting and participating in the ongoing lesson study program in their schools. Administrators also are often directly involved in supporting new teams as they launch their lesson study work. This chapter explores these roles and discusses how teams can partner with administrators to build a strong, sustainable lesson study program in their schools.

In order to provide useful support to teachers engaging in lesson study, most administrators will want to start with understanding why lesson study is important for their teachers and their schools. School principal Lynn Liptak makes a powerful argument for lesson study:

> Like most good investments, we expect that the growth and dividends from the time we invest in lesson study will accrue gradually over a long period of time. Improving our teaching in depth is hard, time-consuming work, which needs to be done collaboratively and in a supportive setting. For too long, professional development time has been allocated to outside experts to "train" teachers rather than given to teachers to reflect collaboratively on their practice. We need to tap outside expertise; we need to improve our content and pedagogical knowledge. But the professional development process needs to occur in the context of our classrooms and be driven as an on-going activity by professional practitioners. (Liptak 2002, 7)

Lesson study can provide many whole-school benefits. It encourages important dialogues about learning and teaching issues that matter to the school community. It also provides opportunities and mechanisms to:

- Work systematically toward improvement in teaching and learning

- Create communication between new and veteran teachers

- Build professional community

- Connect the school to the wider education community

- Increase recognition of the school for its work, through publication, presentation, or hosting of open lessons

Lesson study gives teachers the opportunity to collaborate with colleagues in a professional learning community, and provides opportunities and mechanisms to:

- Address critical problems they face in teaching

- Talk about teaching and learning—with a special focus on students' mathematical thinking and problem solving

- Observe and critique lessons

- Analyze student work with a specific focus in mind

- Adapt instruction using new texts or techniques

- Have their knowledge and expertise utilized and honored

- Build new knowledge and expertise

- Advance professionally by taking on leadership roles, publishing, and presenting their classroom research

A high school department chair and an assistant superintendent, whose districts had been involved in lesson study, reflect on the benefits lesson study provided their mathematics teachers:

I think everyone agreed it was the best PD [professional development] we have ever done. It is too bad not everybody is doing this because you are working with colleagues, talking about content. A lot of people think that is not PD, but it is. And [you are] working on curriculum, you are learning about others who are teaching the same lesson. There is no risk involved; you're working together as a team. It's not just a talk or lecture like other PD. You can't just learn about something in isolation, you have to work as a team and practice it.[1]

[Lesson study teachers have] an openness to feedback, collaboration, and investment in "How do I make this better?" rather than holding responsibility as an individual. How do I make this a better experience for kids, collaboratively? . . . I'm excited about the mix of experience of the teachers, with more and less experience. It's unusual; it breaks down a lot of barriers.[2]

Which School and District Leaders Are Involved in Supporting Lesson Study?

The particular constellation of school and district leaders who are connected with lesson study teams varies greatly from school to school. Active support for the teacher-led

---

1   Interview of high school department chair whose department participated in lesson study (Karp 2005).
2   Interview of assistant superintendent whose district supported a team of high school mathematics teachers in doing lesson study (Karp 2005).

teams often comes from a few individuals whose leadership role supports mathematics teaching and learning.

The principal is the academic leader of a school and as such, is active in thinking about and planning professional development activities for the whole school and in supporting initiatives to improve student learning. For this reason, the school principal is almost always involved with lesson study on some level, and may be the person who first introduces it to the faculty. The school principal is also often the person most able to provide logistical support for the team. This might include permission and coverage for teachers to be released from class for lesson observations or funding for after-school time spent in lesson planning.

Mathematics and pedagogical support for the team are also important and most often come from a mathematics coach or lead teacher, or from a mathematics coordinator or assistant superintendent for curriculum and instruction. These people may be among the first in a district to hear about lesson study and to find funding and professional development to support its introduction. They may also be involved in directly coaching teams and providing content and facilitation expertise. Their role is important also to solidly place the lesson study work of teams as part of larger district mathematics initiatives and to advocate for a long-term commitment to lesson study.

These leaders serve a variety of functions in their schools—from faculty leadership, to supervision, to academic leadership, to community liaison, to administrative organization, to professional learning support, and so on. Lesson study is closely connected to and supported by many of these administrative functions, but at times it requires leaders to adopt a new stance or learn new roles or skills. It is important to separate out the evaluative role (the idea that the lessons are opportunities to critique teachers' skills, that is, supervise or evaluate) and learn how to participate as both a leader and a learner in the new forms of discussion and observation. In their role as supporters of lesson study, leaders may be called on to bring knowledge, new perspectives, implementation support and advocacy, facilitation, and institutional support for the establishment of professional learning communities.

What Does Strong Administrative Support for Lesson Study Look Like?

We can think about the range of possible roles that administrators play in supporting lesson study by looking at four areas. These are based on the critical roles principals play in supporting math program improvement initiatives in general[3]:

---

3 The role of principals in these areas for mathematics programs is outlined in Stimpson (2007). *Evaluation Report of Field Test for Secondary Lenses on Learning: Leadership for Mathematics Education in Middle and High Schools: Abstract*. Newton, MA: Education Development Center, Inc.

1. Authorizing and prioritizing lesson study activities

2. Understanding and engaging in lesson study

3. Relating lesson study work to the wider school and district context

4. Representing and connecting lesson study to a variety of constituencies

1. Authorizing and Prioritizing Lesson Study Activities

One of the most important ways for administrators to be involved in a team's lesson study work is to help create an institutional commitment to lesson study. The fact that an administrator supports lesson study, commits his or her own time to the enterprise, and helps teams secure time for their work sends a strong signal to teachers that the work is of high priority.

> Time is one sure measure of commitment. When teachers see serious time committed to lesson study and the administrators taking time to engage in lesson study, they feel confident of a high level of support for the process on a day-to-day basis and over the long haul. (Liptak 2002, 6)

In the Lesson Study Communities in Secondary Mathematics project, project evaluation data showed that

> time continued to be the most-often cited challenge to optimal team functioning. In fact, almost all of the weakest-functioning teams experienced significant structural impediments related to time. . . . Lack of administrative support, including department chairs, principals, and/or superintendents, was the next most common feature of the weakest teams. . . . A couple of teams lost the support they started the project with, and . . . appeared to suffer as a result; the practice of one formerly very strong team weakened noticeably after several supportive administrators left the system. (Karp 2004)

For lesson study, the gold standard is to have time set aside within the workday for the lesson study work. This assures that lesson study is open to all teachers, and can become part of teachers' regular professional activity. Finding ways to make this happen is challenging, but not impossible if the school administration is committed and creative. Generally, what is needed is roughly an hour per week for two or three months, or some equivalent amount of time in longer meetings. Possibilities include:

- Professional development release days occur several times a year in many schools—offering extended time for research lesson planning.

- Connecting with school improvement initiatives—lesson study can be scheduled as part of the initiative.

- Working with the union. It may be possible to negotiate release from duties or assign lesson study as an academic duty—for collaborative academic work or special academic projects that benefit the school.

- Changing the school schedule. Some schools make the decision to create time within the schedule that is specifically reserved for professional learning.

- In elementary schools: Careful scheduling of "specials" to release grade-level teams.

- In high schools: Converting a portion of department meetings to lesson study.

- In middle schools: Many middle schools are organized around a team structure, with team time set aside for collaborative planning activity, like lesson study.

In addition to helping teams find the time to do lesson study, some specific things that administrators and teachers can do together that will support the development of lesson study at their schools are:

- Form a team of teachers and leaders to attend a lesson study conference or go to a lesson study open house in another district.

- Bring speakers to the school to talk about lesson study.

- Organize an annual "introduction to lesson study workshop" so that new teachers and new team members can learn about lesson study.

- Organize a program to develop local lesson study coaches, who provide ongoing support to teams.

- Form a lesson study planning committee. A small group including teachers, content specialists, and administrators could meet at the beginning and end of the school year (or more often) to organize the upcoming lesson study activity in the school. This might include thinking about which teams are active each semester, deciding whether the school will host a public lesson, organizing coverage in advance, or choosing a common theme to share across all teams.

2. Understanding and Engaging in Lesson Study

I said to teachers, when we do PD, I am side by side with you as learner—not a principal. They see me in that role.[4]

Understanding the benefits of lesson study is a starting point for their support of this work, but administrators can learn much more about the lesson study process and benefit in many of the same ways that teachers do by actually engaging in some part of the

---

4   Loretta Caputo, School Principal, Lesson Study Institute for Coaches and Principals, New York City, May 25, 2006. Statement made during panel discussion of how principals supported lesson study program in their schools.

process. Participating in lesson study meetings or events provides an entry point for understanding and learning. However, when administrators participate in some part of the lesson study process with their teachers, they need to pay careful attention to seeing (and presenting) themselves as learners and supporters, not as evaluators or directors. The purpose of participating is both to learn and to demonstrate commitment to the teachers' work. For administrators, participation opportunities might include:

- *Attending a team meeting.* This would be a good opportunity to learn what the team is researching, see the kind of professional dialogue that goes on, and ask teachers what they feel are the benefits and challenges of the work.

- *Attending a research lesson.* Many teams invite an administrator to attend the teaching of the research lesson and post-lesson discussion. This is a good way to learn about the structure, goals, and protocols of the observation and discussion and to enjoy the opportunity to learn about how students think about the mathematics. Sometimes, the administrator is invited specifically as a special commentator, bringing a new viewpoint or special expertise to the discussion. For a principal, this role may include commenting on how the team's work contributes to school goals. More information about this role is given in Chapter 16, *Incorporating Expertise from Outside the Team.*

- *Joining a team.* In our work, we have seen coaches, department chairs, mathematics coordinators, and assistant principals joining in the work by becoming a full team member, and engaging in all aspects of the work. In most cases, this was welcomed by the team and has been quite successful, but it does require the sensibility as quoted at the beginning of this section. One must participate side by side as a learner.

3. Relating Lesson Study Work to the Wider School and District Context

The principal and other administrators can play a whole-school leadership role by envisioning lesson study as a school or districtwide program. Teachers could meet across disciplines to identify whole-faculty goals for students—goals that address student weaknesses or schoolwide challenges. For example, teachers might set goals relating to improving students' reasoning or communication skills, or to encouraging students to connect their school studies with their lives in the community. Lesson study teams within or across disciplines could then use these as goals in their research lessons, alongside their content-specific goals, allowing the school to make progress on these important issues.

In their role as academic leaders, administrators are often aware of large trends in education that affect many content areas. These trends may suggest common goals that the school faculty can explore through lesson study. For example, a school may be adopting a full-inclusion classroom model in all subjects, and see lesson study as an effective tool for supporting teachers in developing new instructional strategies for this model.

Some specific ways in which administrators can foster a school or districtwide impact involving lesson study are to:

- Provide support for teachers to do lesson study on an ongoing basis, so that their work can solidify and expand.

- Advocate for support from the superintendent for a districtwide program.

- Invite experienced lesson study teachers in the school to present about their work at a faculty meeting, and/or to mentor new teams in another discipline.

- Support whole-school discussion of issues of teaching, problem-solving approaches, and student thinking.

- Establish an annual lesson study conference for the district, at which teachers from many schools would present and discuss their research lessons.

4. Representing and Connecting Lesson Study to a Variety of Constituencies

Lesson study is intensely local and at the same time speaks to a wide professional audience. This is one of its strongest characteristics. Administrators and teachers can foster it as a mechanism for addressing local needs, and they also can partner in finding ways for the team's learning to reach the wider audience through open houses, conference presentations, articles, and the press. In many ways, the principal is uniquely positioned to support impact in both of these arenas.

- Help the team locate and connect with outside expertise. This might involve inviting a speaker to the school for a professional development day on a topic relevant to the lesson study research going on at the school, bringing university faculty or researchers into connection with teams for example. (See Chapter 16, *Incorporating Expertise from Outside the Team* for more detail.)

- Support the team in hosting a public lesson or lesson study open house in which teachers come from other schools to observe and discuss a research lesson. (See Chapter 17, *Public Lessons: The Lesson Study Open House,* for more detail.)

Common Misconceptions That Impact Administrator Involvement in Lesson Study

As teams move forward with their lesson study work and as teachers and administrators think about how to work together to foster successful lesson study practice, keep in mind the following common misconceptions about lesson study:

Misconception #1: Lesson study is primarily a way to generate curriculum or produce a large number of model lessons.

Research lessons are meant to be observed, with the purposes of testing out new ideas and exploring ways to address the important common goals. Over time, many interesting and strong lessons will likely emerge from the testing and improvement process. Other lessons that emerge may not seem dramatically new, improved, or different, yet they will still produce great learning for the teachers and other observers who participate in the process of developing, observing, and discussing them. Research lessons live on through the learning of those involved, through their use by other teachers who adapt them to local student needs, and by providing a link in a larger chain of practitioner research. Each lesson study team may begin a new cycle by seeking prior research lessons to build from.

Misconception #2: When observing the research lesson, the observer's role is to evaluate or supervise the teacher.

Lesson study is not evaluative. Teachers *and* administrators who observe the research lesson (or interact with the team at other times) should see their participation as an opportunity to learn about how students think about and learn mathematics, not as an opportunity to evaluate or supervise the teacher. This is a question of focus and intent. During the post-lesson discussion, for example, everyone focuses more on the lesson design, the content, and the students' interaction with the lesson than on the teacher's skill or pedagogical moves. This does not rule out sharing expertise in a nonevaluative way or discussing the lesson content in depth. Questions under discussion will provide professional learning in many ways—for example, through discussion of: What makes a good understanding goal? What features of a lesson reveal student thinking? What tasks and follow-up questions support learning?

Misconception #3: Lesson study provides a quick way fix to problems in the school.

First, lesson study simply isn't a short-term intervention. It is a model for making steady incremental improvements that accumulate and increase in impact over time. Schools and teams may stay with particular goals for many years. Though there are immediate benefits, it would be unrealistic, for example, to expect test scores to jump twenty points in one year. Lesson study could serve as one powerful element in a program aimed at improving test scores, but its impacts on teachers, teaching, and learning are much broader than those measured by high-stakes assessments.

A second reason why lesson study isn't a quick-fix model is that there is a learning curve for new teams. For most, over the first few cycles, a lot of change takes place in the ability to conduct the work smoothly, and achieve good results. The main idea (conducting a cycle of inquiry) is not complicated, but it does require new skills, new approaches, and a lot of trust and community building. If you find a school that says "We tried it and it didn't work," always ask, "What did you do exactly? And for how long? And with what support?" For an excellent description of one school's long-term commitment to lesson study and its impact on student and teacher learning, see the article by Lewis, Hurd, and O'Connell (2006).

Misconception #4: It is OK to leave out some parts of the cycle.

The reality is that in every school, even within every team, questions arise about adapting the lesson study process to fit the local context. The challenge is to be sure that if any adaptations are made, the essential core elements in the lesson study process are left intact. Some local adaptations remove a truly critical part of the work—for example, having one person teach the lesson and omitting the team observation. Making such an adaptation would severely restrict the amount and variety of data collected, which in turn limits the opportunity to discuss, debate, and learn from the lesson.

Misconception #5: Lesson study is not appropriate for new teachers or for teachers with limited content knowledge.

We have often heard questions like, "How can a team increase its knowledge beyond the level of what teachers already know?" or "Novice teachers don't have much to contribute on pedagogy, and their time is better invested in day-to-day issues." These concerns reflect a misconception that lesson study is only about expert teachers producing expert lessons. These questions may also reveal a lack of understanding about some basic assumptions that drive lesson study and create learning opportunities for teachers at any level of experience or expertise. In lesson study we assume that (1) teachers at all levels own expertise, are familiar with student issues, and are capable of (and invested in) researching ways to address them; (2) incorporating expertise from *outside* the team is a natural part of the process; (3) team members ask for and share knowledge with one another; (4) lesson study opens up doors for learning that did not exist, or were closed, in the prior school culture; and (5) lesson study provides *many* benefits to teachers in areas like leadership development, professional community, pedagogy, retention and renewal, and so on. It is worth noting that some mentoring programs for new teachers include lesson study as a primary component.[5]

---

5   For example, a high school in Illinois requires all new teachers to participate on a lesson study team in their first two years. Also, the Knowles Science Teaching Foundation provides a five-year fellowship program for new mathematics and science teachers that includes a significant lesson study component.

As leaders consider ways to partner with lesson study teams in support of teacher learning, questions the leader or administrator might ask to spark discussion and productive planning include:

- What questions about student learning is your team investigating in this cycle?

- What does the evidence reveal about these questions so far?

- What new questions has this cycle raised for your team?

- What are the implications of this work for your teaching more generally?

- What areas of your team's learning may be important for the other teachers in the school to hear about?

- Based on your new understanding of the students, are there specific goals for students that the entire school might pay attention to?

- What do you need to continue your work?

Recommended Resources

To learn more about lesson study and the administrator's role, we recommend the following readings:

Lewis, C., R. Perry, J. Hurd, and M.P. O'Connell. 2006. "Lesson Study Comes of Age in North America." *Phi Delta Kappan* 88(4): 273–81.

Stepanek, J., M. Leong, and R. Barton. 2008. "Improving Mathematics Through Lesson Study." *Principal's Research Review* 3(6): 1–7.

In this *Leader's Guide*, some resources for administrators include the Phase Overviews for each phase of the lesson study cycle found in Part II; About This Guide, *Developing Lesson Study Practice* (Chapter 1); and other chapters in this section, which include *Planning for Sustainability* (Chapter 14), *Incorporating Expertise from Outside the Team* (Chapter 16), and *Public Lessons* (Chapter 17). Also, the Resources Appendix contains many links and suggested additional readings about lesson study.

CHAPTER 16 | Incorporating Expertise from Outside the Team

A core feature of lesson study is that teachers on the team are the main leaders of the team's work. It is a teacher-driven process that relies on teachers investigating questions that are important to them. However, this does not mean that the work of the team resides totally within the boundaries of that team. The strongest lesson study teams are those that readily share expertise within the team and learn from their own students, while also looking outside the team for fresh perspectives, professional dialogue, and expertise. They define themselves as members of a wider professional learning community.

Outsiders can bring knowledge in areas of mathematics, lesson study, student thinking, or group facilitation, and *as outsiders* are able bring new perspectives, and to notice things the team cannot. The team can also connect to outside knowledge and new perspectives through print resources such as mathematics books, websites, and professional journal articles on teaching and research.

Who Can Best Provide the Expertise the Team Needs?

The individuals who typically serve as *outside advisors*[1] include teachers from outside the team, principals, district curriculum coordinators, mathematics specialists, coaches, other administrators, university mathematics professors, professional development providers, and mathematics education faculty. Other lesson study teams can also be a valuable source of outside expertise. Sometimes, a team invites teachers of other subjects to join the team to offer powerful perspectives that help the team put their lesson in a broader context. Sometimes colleagues are approached because they know about a particular content area (like statistics) that the team is studying, or because they work with particular populations of students (English language learners, special education students, and so on). In other cases, a team is created purposefully to include particular expertise—for example, some schools that want to think about the development of mathematical ideas across the grades bring together a cross-grades team for that purpose. No matter who provides the expertise, we learned in our work that adopting an openness to learning from one another is critically important for outsiders as well as for team members.

---

1   The terms *outside advisor*, *outside expert*, and *knowledgeable other* are all frequently used to describe people from outside the lesson study team who share their expertise, experience, and/or fresh perspective with the lesson study team. These terms are used interchangeably in this chapter.

Ways That Outside Advisors Help Teams

When a team invites an outside expert to join their lesson study process for all or part of a cycle, it is usually to draw on that person's content knowledge, knowledge of lesson study, or their skills in facilitating lesson study discussions. In this chapter, we discuss how advisors contribute in two main realms of expertise that teams seek (lesson study and mathematics or other content), describe a number of specialized roles outside advisors play (coach, special commentator, workshop leader), and conclude with a discussion of the challenges associated with leadership training and development for knowledgeable others.

Contributing Content Knowledge and New Ideas

One of the most common contributions of an outside advisor in lesson study is content knowledge. That content knowledge may include mathematical content expertise, knowledge of student learning of particular mathematics topics, understanding of how a particular group of students learns, familiarity with certain curriculum materials, or background in the use of technology. Outside experts can infuse new ideas and thinking about teaching and learning into a team's work, and improve the mathematical content knowledge of the team. One sign of a team's deepening practice is their reaching out for ways to continue to learn and build a strong mathematical learning community around their work.

- Team A had participated for several years in a research project at a local university about learning algebra in technologically rich mathematics environments. The team developed several research lessons to observe student thinking and learning using a wireless calculator/computer network in the classroom. The university researchers participated in the observation and post-lesson discussion, as well as meeting periodically with the team to discuss lesson design and how the technology supported teaching and impacted student thinking and engagement.

- Team B was composed of two first-year teachers and one uncertified teacher. The school mathematics specialist attended all team meetings. The team spent a lot of time working out the lesson problems together and the teachers turned regularly to the specialist for mathematical (and teaching) support.

- Team C, through grant funding, was able to invite a university mathematician to join their team for an entire cycle. The goals were to deepen team understanding of the lesson topic, enrich the mathematical discussions during the cycle, and introduce the mathematician to the lesson study process. They hoped to form a long-term partnership with the mathematician that would include future work in study groups and research lessons.

- Team D planned a fifth-grade lesson on basic counting principles—How many types of cupcakes can we make if we have vanilla and chocolate cake mix, 3 kinds of frosting, and 2 kinds of sprinkles available? They were curious what features of the lesson design would open learning to the wide range of learners in the class, and so invited the school's special education director to participate in the observation and post-lesson discussion.

There are some teams in which teachers do not have particularly strong mathematical backgrounds. These teams in particular could benefit from additional expertise from outside the team. One way for an expert to support these teams would be to launch a cycle with a workshop around a particular mathematical idea (e.g., multiplication, similarity) and have the team develop their research lesson around that content. Another strategy would be to utilize a professional development resource during the cycle (e.g., a collection of problems, a book such as Driscoll's *Fostering Algebraic Thinking* [1999], or a "Lesson Study Toolkit"[2]).

Contributing Knowledge of Lesson Study

Lesson study is still very new in the United States. For novice teams, it can be helpful to have someone who can explain and model the process, and is familiar with issues new teams face. A more experienced team might seek to improve their practice around some of the more complex parts of the process (topic study, textbook analysis, post-lesson discussions, writing the rationale, etc.). All teams can detect quite readily what the cycle of steps includes, but it is harder to know what the rationale behind the steps or the goals of each process are. This book is intended to provide knowledge in this realm for teams and leaders, and there are a growing number of educators who could contribute this specialized knowledge to teams.

▌ Specialized Roles That Outside Advisors Play

Lesson Study Coach—Advisor During the Cycle

Lesson study coaching is a new role. The coach has a mentoring or guiding relationship with a team, and helps them develop a strong sustainable lesson study practice. To do this effectively, the coach also needs to be knowledgeable about content, teaching, and learning. Many coaches are also skilled facilitators. The coach often serves as a sounding board for the team, offering ideas not only about the lesson study process but contributing questions for the team to consider (e.g., What does it mean to understand this concept?) and ideas and suggestions about the lesson. The coach works hard to understand the team's goals for their lesson, asks questions about the team's current plan, or suggests new ideas for the team to consider. The coach and team might have a full-cycle collaboration, an active email correspondence, or a series of well-timed visits.

---

2   Lewis, C., and R. Perry. *A Resource Guide for Lesson Study on Developing Number Sense for Fractions*, in development, www.lessonresearch.net. Mills College Lesson Study Group.

Many teachers and school site mathematics leaders coach local lesson study teams, but outside advisors are also often tapped to provide coaching. Even if the coach works closely with the team, it is important to also try to bring the fresh perspective of an outsider. (More about the role of a coach or facilitator can be found in Chapter 2, *Team Leadership and Group Norms* and in Chapter 1, *Developing Lesson Study Practice*.)

Commentator at Post-lesson Discussion

A specialized role played by outside experts in the lesson study context is that of the "special commentator" at the post-lesson discussion. This is fairly common at lesson study open houses, where many educators are participating in the observation and discussions. The expert in this case observes the lesson, listens carefully to the comments made by the team and others during the post-lesson discussion, then offers brief remarks that shed new light on the research lesson. This could include putting the discussions into perspective, raising new questions for the team to consider, challenging or confirming the team's hypotheses with data from the lesson, and especially providing new insights that stem from the person's special expertise. The special commentator may participate in the general discussions or invite debate about his or her remarks.

Several examples of this role are detailed in Part IV of this book. In Chapter 21, *The Essence of a Day: An Open House Story,* the commentator was a mathematics education professor and mentor of the team leader, who has written widely on how teachers come to understand students' mathematical thinking. In Chapter 24, *Expanding Lesson Study Practice at Our High School,* the commentator was a university mathematician with a special interest in teaching mathematics through problem solving. He was able to offer insight into the team's strategies for teaching students to use efficient problem-solving strategies.

Workshop Leader—Extending Teachers' Learning from the Lesson

Some lesson study open houses or conferences offer workshops and presentations as well as research lessons. This is an excellent opportunity for someone with the right expertise to lead a discussion or activity that expands on the mathematics of the research lesson. For example, at the 2007 Chicago Lesson Study Conference a presentation on kyouzai kenkyuu was given by Dr. Tad Watanabe that analyzed the student learning trajectory for quadrilaterals in Japanese textbooks, grades 2 through 4.[3] In effect, he demonstrated the lesson study process of text analysis and provided insights on development of early ideas in geometry.

In another example, at the April 2003 Lesson Study Communities Open House, a thought-provoking workshop on using vectors to introduce lines and slope followed the research lesson on linear patters. The selection from the day's program (at the top of the following page) describes its content."

---

3   Tad Watanabe, Kennesaw State University, Department of Mathematics.

Slope—A Radical Proposal[4]

Is there a way to teach slope that is more closely aligned with students' natural ways of seeing and thinking about lines? In this session we will consider the more radical possibility of grounding the Algebra 1 study of linear equations and slope on the use of vectors. One needs very, very little knowledge of vectors in order to do this successfully. After hearing an explanation of the "radical proposal" teachers will work on sample problems that follow this new approach. The session will include time for discussion: Do these ideas seem more natural? What are the potential benefits? Could they be used alongside a more traditional approach?

Finding and Integrating Outside Experts

The team will need to ask themselves what kind of expertise they need and who might be a good candidate for providing that expertise. If the team begins by considering the categories of expertise and knowledge described in this chapter (that is, content expertise and experience with lesson study and facilitation), they may be able to identify some needs. Furthermore, administrators and leaders in the district may know people who could fulfill these roles. Over time, districts should work to develop leadership potential within their district by helping individuals take on the role of outside advisor and by forming connections with local universities so that they can serve as a source for knowledgeable others.

Preparing Outside Experts

Individuals with expertise in mathematics or other specialties may be called on to play the role of *outside advisor* to lesson study teams, even when they themselves have had little experience with lesson study. Similarly, teachers and other educators may have been involved in doing lesson study, yet never have played particular leadership roles—like coach, or special commentator. A school principal, who may have plenty of leadership experience, could be asked to be a special commentator at a research lesson when she has never attended one. In all of these cases, some preparation or training, as well as advance planning by the team, can be very helpful.

For outside advisors who are new to lesson study, here are a few ideas to consider. Attending a lesson study open house[5] is one way to gain a sense of the nature of

---

4   Workshop led by Joseph T. Leverich, mathematician, teacher, lesson study coach, Lesson Study Communities in Secondary Mathematics Open House, March 3, 2004, Foxborough, MA.

5   For more information about open houses see Chapter 17. *Note:* The Chicago Lesson Study Research Group (www.lessonstudygroup.net) sponsors an annual conference in an open house format.

lesson study research. In this setting, one sees the kinds of questions about practice that teachers investigate through lesson study, participates in a post-lesson discussion, and observes how data on students' thinking contribute to the team's understanding. Another way to introduce outsiders to lesson study is to have them visit a lesson study team meeting as an observer. Written descriptions of lesson study[6] and its goals, or introductory videos are also available. The goals of these introductions would be for the outside advisors to gain understanding of how lesson study works to improve student and teacher learning, to become aware of the content and mathematical questions that the team is investigating, and to think about the learning potential of the collaboration for all parties.

For lesson study coaches and knowledgeable others, the issue of professional learning opportunities for leaders is important. Ways to deepen the leaders' practice, ways to train new coaches for teams in a district where the practice is spreading, and ways to nurture teacher leadership are all necessary. Tad Watanabe summarizes this issue well.

> Although classroom teachers are the main players in lesson study, all knowledgeable others play important supporting roles in improving the quality of lesson study. However, just as teachers must continuously work on improving their craft of teaching, knowledgeable others must also strive to improve their crafts. Just as teachers can learn so much from their own students, knowledgeable others can learn from teacher practitioners of lesson study (and students). Perhaps such a learning stance is the most important characteristic of an effective knowledgeable other. (Watanabe 2005)

Conclusion

As teams work to sustain their lesson study practice and make it stronger, it is important for them to think about how to integrate a variety of perspectives into their work and to experiment with different ways to learn from outside experts. Some questions for leaders and teams to consider when reflecting on how they are making use of outside expertise include:

- What have we learned from someone outside of the team during recent cycles?

- How could we stretch ourselves mathematically? Who could help us do that?

- How do we communicate our goals and theories to invited guests so that they will be able to provide us the best feedback and push us with the most challenging questions?

---

6   Chapter 1 would serve this purpose, as well as many of the articles noted in Resources Appendix.

- In what ways have we drawn on outside resources during each part of the cycle: when setting goals, when developing the lesson, when observing and discussing the lesson, and when analyzing what we have learned at the end of the cycle?

Resources

There are a number of other useful resources with information and examples of the use of outside expertise in lesson study.

Chokshi, S., and C. Fernandez. 2004. "Challenges to Importing Japanese Lesson Study: Concerns, Misconceptions, and Nuances." *Phi Delta Kappan* www.pdkintl .org/kappan/k0403cho.htm.

Lewis, C. 2002. *Lesson Study: A Handbook for Teacher-Led Improvement of Instruction*. Philadelphia, PA: Research for Better Schools.

Lewis, C., and R. Perry. *A Resource Guide for Lesson Study on Developing Number Sense for Fractions*, in development, www.lessonresearch.net. Mills College Lesson Study Group.

Wang-Iverson, P., and M. Yoshida, eds. 2005. *Building Our Understanding of Lesson Study*. Philadelphia, PA: Research for Better Schools.

Watanabe, T., and P. Wang–Iverson. 2002. Role of Knowledgeable Others, presentation at the Lesson Study Conference, Stamford, CT, November 20–22. See www .rbs.org/lesson_study/conference/2002/papers/watanabe.php.

CHAPTER 17

Public Lessons: The Lesson Study Open House

What Is a Lesson Study Open House?

Teachers who are doing lesson study almost always invite a few colleagues to attend the observation and discussion of their research lessons in order to share what they are learning and bring new insights to the team. Sometimes the lesson study team decides to share their work more broadly by inviting more than one or two colleagues—perhaps even inviting educators from beyond their school—to observe and discuss lessons in an *open house* or *public lesson*[1] format.

The *lesson study open house* is a teacher-led day of professional learning, at which observation and discussion of one or more research lessons at a host-school is the focus of the day. A diverse group of educators may come to the open house—teachers from the host school and nearby schools, administrators, and special guests who contribute particular expertise on the team's research focus could all be invited. All attend because it is a great opportunity to learn—by seeing actual classroom lessons, observing how students grapple with the mathematics, hearing what the planning team expected to happen and why, and discussing issues of teaching and learning that arise. The open house takes the work of individual teams to a deeper level and allows the team's research to have a wider professional impact.

Why Might a Team Decide to Have an Open House?

The potential benefits of an open house are numerous, but the decision to host one is usually based on the lesson study team's goals. These team examples may help your team think about various reasons for having an open house.

- *A standard team practice.* Some teams routinely open up their research lessons to outside observers—by inviting guest educators to the observation and post-lesson discussion in almost every cycle, or by hosting an annual open house.

- *Culminating a period of work.* A team that has stayed with the same research theme or content focus for several cycles of lesson study may want to culminate the work of investigating that research theme or content focus by

---

1 The term *public research lesson* is often used interchangeably with *open house* to refer to research lessons that are open to a wider audience or are hosted at a public event, like a conference. In this chapter, we focus largely on *school-hosted* public lessons and use the term *open house*.

consolidating and sharing their learning about it from several cycles, and see the open house as a way to bring this learning to a larger group of colleagues.

- *Deepen learning in a single cycle.* A team may have chosen a particularly difficult or important content focus or goal for their cycle, and choose to host an open house to deepen their understanding by gathering input from guest educators during an open house.

- *Multiple team sharing.* Several schools in one district or several departments in one school often share the same broad goals for students, and find it rewarding to have a lesson study open house as a way to build common understandings and make progress on shared goals.

- *Initiatives that go beyond the team.* A team may be using their lesson study work to support a schoolwide initiative, such as implementation of a new curriculum with a totally new pedagogical model, or efforts to improve achievement among underserved populations. The open house can be an effective way to share and build understanding around the goals of this initiative.

- *Building a lesson study program at your school.* Inviting colleagues from your own school to an open house is a powerful way to show them what lesson study is all about. By participating, teachers can experience the nature of the professional discussions and learning involved in lesson study and see what the team's work has produced. This learning can serve as the starting point for integrating new teachers into lesson study teams or starting additional teams at the school.

For each of the reasons for holding an open house listed here, the team is focused on some combination of sharing what they are learning with a wider educational community and deepening their own learning through the insight of the wider educational community. The two-way sharing of learning between the team and guests invited to the open house make the team's efforts part of a larger stream of professional learning.

What Happens at an Open House?

Lesson study open houses can be simple or complex—involving a small or large number of participants and one or several research lessons. Have you seen images of Japanese research lessons that involved the whole faculty of a school, or of public lessons at national conferences with hundreds of teachers observing the lesson in an auditorium? Eventually, these models may become more common in the United States, but smaller varieties are equally worthwhile and may offer some benefits that the larger open houses cannot. When a team is hosting an open house at their school, it has to be on a scale that the team is comfortable with and that works logistically at that school. Based on our experience, an open house in which ten to twenty-five teachers observe any lesson that is taught during the day is an excellent place to start. This size allows the observation to take place within a large classroom, enables all observers to be close enough to see and hear students well, provides all observers a chance to participate in the post-lesson discussion, and brings a rich variety of expertise and viewpoints to the discussions. It is also easier to organize than a larger open house. As teams gain experience with this model, there are many benefits to going bigger at some point, but the reasons for doing so should emerge naturally from the work of the teams.

Who Attends?

The *lesson study team* that planned the research lesson invites a range of educators to be *participants* in the activities of the open house day. The participants are usually teachers, but the group also includes others from the wider educational context—perhaps university faculty, mathematicians, coaches or specialists, school administrators, and people with expertise related to the team's work (for example, special educators, curriculum specialists). The team thinks carefully about who they would like to invite, trying to include a variety of people who will provide rich and varied expertise. Often, the team invites one person with special expertise to offer comments at the end of the post-lesson discussion. This commentator makes remarks aimed at taking the discussion to a new level, offering summative insights, or suggesting new directions for research on the topic. (See Chapter 16 for more information about this role.)

What Events Are Scheduled?

Most open houses take place at a school and include these basic elements:

- *Welcome.* Attendees gather, meet each other, are welcomed to the host school, perhaps with refreshments, and learn about the day's events. The principal, superintendent, team member, or mathematics department chair may make these welcoming comments.

- *Introduction to the team's goals and lesson.* To learn from the lesson observation and contribute to the post-lesson discussions, participants need to know something about the team's goals and research questions, what the mathematics content of the lesson is, and what observation questions the team is seeking to answer. Participants should *read the lesson plan* and to *do the key mathematics problems* in the lesson. Usually the team presents this introduction, or plans introductory activities that allow participants to understand what the team's research is about and to focus on how the lesson helps students understand the mathematics. The team shares its observation questions and any "rules" they wish observers to follow (e.g., don't talk to the students during the lesson, sit at these chairs at the tables while students are working so you can see their work, refrain from side conversations with other observers during the lesson).

- *Observation of the research lesson.* In a small open house, the research lesson is usually taught in the classroom of one of the team members to a class of students that the team member regularly teaches. All observers take detailed observation notes on the lesson with a strong focus on how students are thinking about and understanding the mathematics. The goal of the observation is the same as it would be for any research lesson: to provide content/data for an evidence-based discussion of issues of student learning. Depending on the number of observers, some limits may be placed on where they sit and stand. In a large open house, the lesson may take place in a library or other large school space and the team will have to plan ahead to make student voices heard and student work visible.

- *Post-lesson discussion and reflections.* The post-lesson discussion has the same goals as a team post-lesson discussion (refer back to Chapter 10 for more information), but with many guests may be more formal and usually has a moderator who describes the format up front and then facilitates the sharing of data and discussion.

- *Closing.* At the end of the day, participants often gather to share final reflections, or for refreshments and informal conversation or networking.

Many teams add to this core set of open house activities by planning extra sessions related to the team's overall goals for the day. Even if the open house is happening within regular school hours, there is usually room for at least one additional element. Sometimes, additional activities are aimed at sharing about the lesson study process. For example, the team might provide an introduction to lesson study that allows those participants who are new to lesson study to understand the context of the team's work and to participate effectively. In other instances, the extra element is focused on extending learning about the team's research theme or mathematical topic. A team might include a student work analysis session in the day, or invite a guest speaker to lead a workshop related to the mathematical content of the lesson, the lesson pedagogy, or a school initiative (see Chapter 16 for more information). For example, in a mathematics workshop at an open house, participants may be asked to do mathematics together to extend their understanding of the topic and of its mathematical connections. In cases where multiple teams at the same school or across schools have been exploring the same topic or research theme, cross-team sharing about what has been learned is a natural fit for the open house setting. Generating more opportunities to observe lessons is another example of how teams make use of the open house. Some teams include more than one research lesson observation during the open house day.

The team might invite participants to observe more than one lesson, or ask participants to choose among multiple research lessons that are taught concurrently. Some teams even schedule in some time during the day for observation of regular classroom lessons at the host school, offering an "open-door" model during which guests can drop in to observe part or all of a regular classroom lesson.

Practical Planning

Reading about all the possibilities for the open house day may seem a bit daunting, but it is important to remember that each open house is planned by teachers to be doable in their school setting and to address questions that are important to them. It can be quite modest in scale if the team wishes it to be. The major work involved in preparing for any open house is what a team always does in their lesson study cycle—that is, develop the research lesson. However, the team will need to have the open house in the back of their minds at particular points during the development of the lesson. In addition, the team will need to do some envisioning of the open house and will need to plan for logistics. All of these facets of planning for an open house are described next.

Developing the Research Lesson That Will be Taught at an Open House

Your team will develop the research lesson during a regular cycle of lesson study, using your normal lesson study process and asking all the same questions you would usually consider, including: What do we want our students to understand about this mathematics? What kind of methods and thinking do we expect from our students on these problems? How exactly does our lesson help achieve our broad goals? The team will also need to try to address some issues that relate to the open house setting when developing the lesson, including:

- *Owning the lesson ideas.* Your team will need to talk through the lesson rationale and design thoroughly so that the whole team is solidly grounded and together on what you are trying to accomplish in the lesson and why the team made its choices. Full team ownership of the ideas means the team can respond well to questions and ideas that surface at the post-lesson discussion during the open house and raises the overall level of discussion.

- *Getting grounded in the mathematics.* The better grounded your team is in the mathematics, the better they will be able to learn from and dialogue about observations and ideas that come up in the lesson and at the post-lesson discussion.

- *Choosing a good topic.* Your team will immerse itself in this topic quite deeply—so finding a topic and research question that the team truly is interested in exploring is important. Choosing a topic that matters mathematically will allow the open house to impact teaching in the many classrooms represented by not only the team members but all of the invited guests.

- *Considering how observers will be able to see and hear students doing and talking about the mathematics.* When there are many observers at the teaching of the research lesson it can be challenging to allow access to what students are doing and saying for all observers. Planning a lesson that requires students to solve a challenging problem, so that they have something to think about, and that includes ways for students to make their ideas public, is therefore particularly important for an open house lesson.

Envisioning the Open House

Your team will need to think together about your goals for the open house and will need to sketch out a broad framework for the open house. This overall planning does not need to happen at the same time as the development of the lesson. It can happen at the beginning of a cycle of lesson study, or as part of your team's long-term planning—perhaps in the spring when the team is reflecting on its goals and making plans for the following school year. Questions the team may ask while envisioning the open house include:

- *Why do we want to have an open house?* What do we want to share about our work? What do we want to learn? Who else might benefit from our research? Can the event introduce our school colleagues to lesson study?

- *What events will our open house include?* How many people? How many lessons? What other activities will support our broad goals?

- *Who might support our goals and help us prepare?* Who might lend practical help, space, time, funds, and food? How can we plan to create minimal disruptions of the school day?

Logistical Preparation

Key logistical planning for an open house includes inviting guests, creating handouts for invited guests, finding and reserving meeting space, obtaining any needed permissions (e.g., if videotaping is involved), making arrangements for refreshments and parking, and considering other logistics and scheduling for the open house day. Questions the team will be thinking about include:

- *Who will we invite?*

- *When will the open house take place?* How long do we need to plan our lesson and is there enough time between now and the proposed open house date? How can we plan the schedule for the day to mesh smoothly with the school day?

Will the lesson happen during regular class time? Will guests arrive at a time that will avoid student bus arrivals?

- *Where will all the events take place?* What rooms will work for the open house while dislodging as few colleagues as possible?

- *How will we prepare the students who will be observed?* Most students are willing to participate, once they understand what it is about, and many are even excited or honored that their class was chosen. The team needs to have a plan for how the teacher will explain the purposes and nature of the lesson observation to the students and address any concerns.

- *What materials do we want to provide for all the participants?* For all open houses, basic lesson handouts and lesson plans are the primary materials needed. Participants will also need access to a detailed lesson plan that explains the team's rationale, plan for teaching, and observation questions. Giving participants a book or article about lesson study or about the mathematical focus of the lesson, and an open house booklet with all of the day's materials adds to the professional impact of the day.

- *How will the day's events be recorded?* Video has major benefits for recording the discussions and learning of the day but it also creates challenges related to devoting time and energy to setting up the technology, obtaining permissions, and in some cases changing the nature of discussions. Other records of the day that your team might want to preserve include student work, notes from the post-lesson discussion and from presentations by speakers, and written end-of-day reflections or evaluations from participants.

Traditions That Matter

If you have not attended a lesson study open house before, it might be hard to picture the tone and culture of these events. We can share some of the traditions or protocols that we have observed at open houses that contribute to their professionalism and impact.

We, Not *I*

The teacher of the lesson is just one member of the team—and the lesson that is observed and debriefed represents the ideas and work of the full team. Questions and comments are addressed to the team as a whole.

Whatever Your Role, You Are Here as a Learner

Team member, mathematician, school principal, student, teacher, coach—all have an opportunity provided by the team to discuss issues that are raised by the lesson.

To emphasize this, observers often thank the team for this opportunity to learn, and share something that they learned when they speak at the post-lesson discussion.

Evidence-based Comments

Open and evidence-based discussion is expected in the post-lesson discussion—and this can include both positive comments or praise for the lesson and challenging questions about the lesson design and decisions the team made in planning. Always, however, it is expected that observers will base their comments on specific observation data.

Discussion of Data Before Revision

One can assume that the team has put a great deal of thought into the way the lesson was planned. At the post-lesson discussion, participants usually focus on sharing their observations with the team, asking the team to share their rationale for choices in the lesson, and working together with the team to try to understand any surprises that occurred in the lesson. The team will meet later to think about how all of these ideas lead to productive revisions in the lesson.

Celebrating the Team's Hard Work

When the day is over, it is traditional for the team to celebrate, perhaps by going out for dinner together. Special guests to the open house might also come to this celebration. During the open house, discussions will have focused on student learning, and serious critique and debate of educational ideas are the order of the day. But at the end of the day, the team will want to thank everyone for the role they played in making the open house a success, with particular gratitude to the team member who taught the lesson. Regardless of how well the whole team planned the research lesson, it is still a big challenge to teach in front of many observers. Mostly, this celebration is just a nice way to relax, enjoy the fact that the team's hard work paid off, and recognize the team for this major professional contribution.

In Summary

As you review this list of protocols and traditions for open houses, you have probably noticed that the overall culture of lesson study open houses fosters an environment that deepens the learning of all who are involved and that is respectful and celebratory. For whatever reason your team chooses to host an open house, and regardless of which activities your team chooses to include in the open house day, if your team and invited guests can participate in a day of learning and sharing of knowledge then you will have made an important contribution to the education profession while gaining deeper insight into your own team's work. (For additional images of lesson study team's open houses, refer to Chapters 21 and 25.)

Envisioning Lesson Study Implementation and Practice

Part IV contains chapters written by members of the lesson study teams that we have worked with over the past several years, primarily as part of the National Science Foundation-funded *Lesson Study Communities in Secondary Mathematics* project. These chapters provide images from actual school teams of various aspects of lesson study team practice including goal setting, lesson development, and the preparation of a public lesson for observation and discussion. Some of the authors tell about the work of a team within a particular cycle, while others speak to the evolution of lesson study within a school or district. Most of the chapters are written by teachers; a few of them come from the perspective of a coach or a school administrator. Each story represents the author's perspective and is not intended to represent the perspective of others from the same school or district.

These teams and their stories have provided much inspiration for our work and we hope that they will do the same for you. You can use these stories to learn more about particular elements of lesson study, or to gain a better understanding of team implementation of lesson study. The authors share the challenges and successes in their lesson study practice. You can read the chapters individually or as a set. Some chapters illustrate key ideas of the lesson study process. The stories included are:

- *Chapter 18: The Importance of Goals in Lesson Study*—A middle school teacher describes how her team set goals together and how those goals informed their ongoing lesson study work.

- *Chapter 19: Josh and Betty's Lesson Study Journal*—Excerpts from two high school mathematics teachers' journals describe their first cycle of lesson study and the development of a lesson focused on the triangle congruence postulates.

- *Chapter 20: How Lesson Study Changed Our Vision of Good Teaching*—This chapter captures the development of a learning community within a cross-grades (grades 7–12) mathematics team, and describes how their thinking about teaching and student learning in mathematics is influenced by their work together.

- *Chapter 21: The Essence of a Day: An Open House Story*—The experience of an interdisciplinary team at an agricultural high school on the day of their lesson study open house is described by a member of that team. Two related lessons were included in that open house, one investigating whether quadrilaterals tile the plane and the other a woodshop lesson using the same mathematics ideas to build a stool.

- *Chapter 22: The Longmeadow Story: A District Lesson Study Initiative*—A coach shares her perspective on working with an elementary district that embraces lesson study as a way to enhance mathematics teaching across the district's three K–5 schools. The story includes a description of the district's experience with using vertical teams in their lesson study work.

- *Chapter 23: Our Lesson Study Journey at King's Highway Elementary*—An assistant principal tells about her own learning about lesson study and how she introduced it to teachers at her school. She shares how a third-grade team at her school used lesson study to address an identified area of weakness on their state mathematics assessment.

- *Chapter 24: Expanding Lesson Study Practice at Our High School*—This chapter chronicles the evolution of lesson study within a suburban high school from a single lesson study team to include multiple teams of teachers from other disciplines, and the growth of teacher leadership as members of a lesson study team become coaches for their colleagues.

- *Chapter 25: Our First Open House: Exploring Spherical Geometry*—This chapter gives an overview of a team's efforts to plan and organize a lesson study open house featuring a research lesson designed to engage students in non-Euclidean geometry.

A set of discussion questions is offered at the end of each team's story in Part IV. The discussion questions can be used for individual reflection, or to prompt discussion about aspects of lesson study practice (e. g., goal setting, hosting a lesson study open house) among a group of teachers. These questions are an additional resource that you may find helpful in reflecting on activities or decisions made in the team stories or in connecting to your own lesson study practice.

| # The Importance of Goals in Lesson Study

Andrea Plate

Lesson study began in our suburban district when I learned of the EDC Lesson Study Communities in Secondary Mathematics (LSCSM) project. After garnering administrative support from the district to develop a team, I sent out an email to the mathematics teachers, and six teachers volunteered to participate. Our first lesson study team included sixth-, seventh-, and eighth-grade math teachers, and the math coordinator. Our after-school schedules had already been set and conflicted with each other, so we chose to meet at 7:00 AM once a week.

During that first cycle we tried to follow the model of lesson study as we had learned it from the LSCSM project workshops and from our readings. We wrote team norms with the goal of having a nonjudgmental team in which everyone could contribute while staying on task. We assigned jobs on a rotating schedule. There was a lot of conversation about how jobs should be rotated with everyone making suggestions for improving the procedure.

Setting an overarching goal was critical. We knew why we were there—to be the best teachers we could possibly be—so this shouldn't have been difficult, right? In retrospect, it makes sense that if working out a system for assigning jobs required a lot of conversation that first meeting, setting an overarching goal would require much more time!

At first we thought our goal was the mathematics. So, we framed content as our overarching goal: "Students should understand algebraic reasoning." But then we debated, Why algebraic reasoning? What about geometry? Statistics? All the other content we needed to include?

Then we moved onto other questions about an overarching goal. What does it mean to understand something? Maybe *understand* isn't the best word. How about *master* or *apply*? We asked each other questions: If it's an overarching goal, how do we know we're moving toward it? How do we measure it? How do we *say* it in a succinct sentence or two?

Slowly we came to realize that our overarching goal wasn't just the mathematics, it was the mathematical student. We talked about what we wanted our mathematics students to look like: We wanted students who were curious, enthusiastic, who were willing to attack problems that were outside their experience, who were creative in their efforts, and who applied what they knew.

We realized that we were talking mostly about problem solving. But what about skills mastery? Isn't it important for students to master the skills of mathematics? Yes, of course this was expected.

What about the student who, because of personal limitations, may not reach that lofty goal of "independent creative problem solver"? This led to discussions about the goal of differentiated instruction. Were we guilty of low expectations of some students? Shouldn't we still expect all students to be willing to approach a new problem with enthusiasm? Again, we challenged each others' foundational beliefs.

We fussed over the exact wording and the length of our overarching goal, but the more we talked, the more important it was to us to say what we meant precisely. After all the debates and discussions, our goal became "to develop a mathematics student who is curious, takes risks, has skills mastery, and is an independent thinker who is capable and willing to approach new situations." Now we were ready to move on.

Looking back, I realize that the thinking we did together to establish our goal in that first cycle affected every research lesson we developed. Let me give you an example. In one cycle we decided to focus on a weakness we had seen in how our students approached math word problems. Too many students gave up, grabbed some numbers, and performed an operation. Or, they found an answer and did not evaluate its reasonableness, sometimes coming up with absurd answers. Every year, the curriculum taught problem-solving strategies, but students applied those strategies or checked their answers only when they were prompted.

We decided to look for some good problems to use as the foundation of the research lesson. For several weeks, our assignment was to bring some good word problems to our meeting at which time we did the problems together and talked together about what prior knowledge students needed to be successful. Many problems that were brought to the team looked like this one:

> *In science class you have six 100-point tests. You need an average of at least 90 points to get an A. Your scores for five 100-point tests are 88, 92, 87, 98, and 81. What is the lowest score you must get on your sixth test to still get an A in the class?*
>
>   *a. Define your variable. (HINT: You're looking for the sixth test score.)*
>   *b. What is the least number of total points you must have to receive an A?*
>   *c. Write the equation for the sum of the six tests.*
>   *d. Solve to find the lowest score you can get on the final test to get an A.*

We were all very familiar with this format, but something felt wrong as we looked at the problems that we brought to the meeting. After several weeks of doing these problems together we finally realized that the student did not need to think through the problem but only needed to "follow the steps" they were given. We were not developing an independent thinker as our goal stated. By just deleting the directions listed in a, b, c, and d, we could change the job of the student completely. This was a major aha moment for all of us and we began to look at what *we were asking the student to do* when we taught problem solving.

We also paid more attention to *what kind of thinking the problem required of students*. By eliminating the *a*, *b*, *c*, and *d* directions, we would be setting up a situation where students would have to think about what the question is asking, what an average is, and come up with mathematical strategy for solving the question. If reading carefully, students would also have to convince themselves that they had found the *lowest* score.

We became more aware of the *variety of approaches* a student could use to approach the problem. By eliminating direction *a*, we opened up the problem for students to start solving it in a number of ways, not simply by defining a variable. Even though it is more challenging overall, it is more accessible. The first step might be drawing five test papers with scores and one test paper with none. Algebraic solutions might emerge from an arithmetic beginning point.

Our new approach to choosing appropriate problems raised the level of thinking we had to do as a team. For example, we had to ask ourselves what student understandings are we really seeking? What problem-solving strategies will students be most likely to follow? If we encourage students to follow different approaches, how will that help them build these understandings? How will we reinforce independent thinking if students are totally stuck?

Taking this approach may not have been the most efficient for us in getting our lessons developed. At times, we became frustrated when it seemed we hadn't accomplished much in a meeting. Yet, years later these conversations continue to mold our instruction not only in a research lesson, but in all lessons. When planning lessons, we ask ourselves: "How do I teach this mathematical idea in such a way that I am also developing an independent mathematics thinker?" I also believe that we established the groundwork for the shift in school culture that followed over subsequent years from a skills-based school toward a more comprehension-based general program.

▌ Discussion Questions

- How did your team choose your broad goals? When (or how) have your goals played an important role in your work?

- The author says, "Slowly we came to realize that our overarching goal wasn't just the mathematics, it was the mathematical student." What do you think this meant to the team? How does this statement relate to your thinking about goals?

- For this team, the goal of fostering independent thinking over time became an ongoing guiding question in their thinking about teaching. Which questions that your team discusses might have a similar ongoing impact on your thinking about lessons or your teaching?

CHAPTER 19

Josh and Betty's Lesson Study Journal

Joshua Paris and Betty Strong

Josh Paris and Betty Strong worked together as a lesson study team in the Lesson Study Communities in Secondary Mathematics project from 2002 to 2004 and completed four research lessons together for a freshman geometry course—Math I Intensive. They teach at a diverse, highly academic, historically innovative high school of 2,000 students. Another small team from the mathematics department also participated in the project and the two teams joined together for the observations and postlesson discussions of their research lessons. Josh and Betty kept reflective journals in the fall of 2002 while working on their first research lesson.

Betty's Lesson Study Journal—First Entry

Josh and I teach a course together—Math I Intensive—which is a standard-level geometry course. It is designed for students who have not experienced tremendous success in a fast-paced traditional math class setting (like Algebra 1 in the eighth grade). Our course is grounded in "understanding." It is very important to us that students have the opportunity to make their own meaning of the relationships in geometry, and not just be lectured to and asked to memorize. Josh and I are both very invested in this idea, and have spent considerable time modifying the curriculum and creating materials for this course. It is our ultimate goal to design a "foundation lesson" for each topic that will really cement an understanding of the big ideas of that topic in the minds of our students.

September 10: Josh's Journal

After attending the two-day lesson study workshop I have a lot to think about. Since Betty wasn't able to attend, I just came up with some rough ideas about what we want to accomplish in teaching Math I Intensive. Here are some thoughts. I think that we want to give our students a sense of self-confidence. Many of them have had difficulties being successful in other math classes. If they can have a positive experience during their freshman year, then hopefully they can build on that and be successful throughout high school and college.

I want them to be good thinkers. I want to help them develop their reasoning skills, both deductive and inductive, so that they can reason through a problem, mathematical or otherwise, successfully. I want them to be able to learn on their own. We also focus on the details in Math I Intensive: notebook, homework completion, organizational skills, note-taking skills. These will all help them to be lifelong learners.

September 25: Betty's Journal

We started discussing a bunch of business about meeting dates, materials, and so on. Josh and I are using an article from the *Mathematics Teacher*[1] magazine as the basis for our lesson. Last spring, when that particular issue came out, we both read the article and commented to each other how it would be a perfect lesson idea for our standard-level Math I Intensive course, but wondered who has time to make it up? It's materials and planning heavy. It's not the kind of lesson you can just whip together. The article is about a lesson to introduce the triangle congruence postulates. It is a game in which students have to "guess the mystery triangle." Kind of like twenty questions. The teacher knows what the "mystery triangle" is and students have to gather information about the triangle by asking questions and use that information in order to draw the triangle. Today, we went back and underlined things in the article we liked, or parts that we thought wouldn't quite work. In this first meeting we mostly discussed the lesson in a general way and took care of the project logistics.

I think our group is unique, because Josh and I already work collaboratively a lot on this course. To me, it is just great to be given this time to put together a very carefully thought out, effective, and meaningful lesson. I think both Josh and I have the same long-term goals for students, which to me is for them to develop a lasting understanding of the basic fundamental relationships of geometry, and to give students ample opportunities to gain that understanding.

October 1: Josh's Journal

Betty and I have decided our lesson will be on congruent triangles. Why congruent triangles? I've never been completely satisfied with teaching congruent triangles. The students never understand them the way I want them to. They understand what they are, but then when we move on to trying to prove them to be congruent, they get confused. Students become even more confused when we then begin to talk about corresponding parts.

---

1   "Letting the Cat Out of the Bag. . . . To Make Room for a Triangle" by Marie Reynolds, *Mathematics Teacher*, January 2002, *95*(1): 6–7. In the lesson described, the teacher picks a "mystery triangle" out of a hat. Students ask questions about the mystery triangle and use this information to draw a congruent triangle. Over time, students get better at asking questions, discover cases in which three questions are sufficient, leading to the triangle congruence postulates.

Furthermore, why do we teach them? Why are they such a big part of high school geometry curriculum? I think that studying congruent triangles is the means by which we can teach students logical reasoning. We study them so that they have the chance to prove some mathematics. But, why is this important? Well, the reasoning needed to go from having certain information, deciding if they have enough information to prove that triangles are congruent, and then using the fact that they are congruent to find additional information (corresponding parts) is a complex process, a process that develops logical reasoning. However, ultimately, who cares? Rather than focusing on just proving corresponding parts are congruent, our goal is to show them a reason *why* it is important. That is, we will be using congruent triangles to prove certain properties of the special quadrilaterals. In that way, the students will see a use for the math that they are hopefully learning.

Also, instead of focusing on rigid two-column proofs, we will be looking more at paragraph proofs. That way, we will be working on writing skills as well and at forming cohesive arguments within a paragraph, skills that transfer to other classes.

October 2: Betty's Journal

Today, we started talking about the specifics of the lesson we're going to teach, and its place in the curriculum topic. We decided to use this lesson as our introductory lesson on congruent triangles. We debated about whether we should first go over some vocabulary, like *corresponding parts*, or explain what it means for two triangles to be congruent. But we agreed, after some debate, that it would be best to save the "names" of things until after the foundation for these concepts is laid. We were also concerned with how much time this lesson could take up.

We began planning the lesson as if we would just describe the rules of the game and then let the groups of students start playing it. In both of our classes, students sit in groups of three or four students, so they would play as a group. The teacher would pick a "mystery triangle" out of a hat, then go around and answer students' questions (like "What is the length of side AB?") and students would start drawing their guesses.

We soon realized that we needed to sit back and consider each moment, imagine how it would go, and plan and adjust this lesson accordingly. Once we started moving through the lesson moment by moment, we quickly realized that our students would probably not understand the rules and would be very needy and lost. So we decided it would be a good idea to play one game together as a class. Hmmm.. . . this is getting tricky.

October 8: Josh's Journal

Well, we've been working on developing the Triangle Mystery Game. Students have to draw a triangle that matches a mystery triangle by asking the teacher questions similar to the way *20 Questions* is played. I think this will be really successful. The key to congruent triangles is that you only need to know three pieces of information in order to know that triangles are congruent. But these students never seem to get there without me just telling them. This activity will be different. If all goes well, *students* will discover that three pieces of information is enough, and will also discover which three are needed.

This seems to be the reverse of what I have done in the past. I normally started with one piece of information, showed the students that this wasn't enough, moved to two pieces, and then to three. This game, however, works backward. At first, students ask more than three questions to guess the triangle and we ask them, "Do you think you can do it in fewer guesses? How many guesses is the minimum?" We are doing this game before we even mention congruent triangles. This is our strategy throughout the course. We always focus on the mathematical relationships that we are working with before giving them names. After all, the only reason certain objects have specific names is because there is something special about them. Why name them before finding out what is special about them?

October 23: Betty's Journal

Now we're into the details of the lesson. Our lesson is more than a one-day lesson, so we've been discussing how we'll partition it up with appropriate homework assignments in between. We've made a conscious decision not to "teach" any new vocabulary or ideas before we start playing the Mystery Triangle Game. We'll just use the words that the class knows and let ideas come out as kids have them. We decided that we'll spend almost an entire day learning the rules of the game and playing a practice round as a class.

We predict that students will have a lot of questions about how to play. We decided that on that first day, we will purposely choose a student to share a question that is off target—for example the student might ask, "Is it an isosceles triangle?" This is not the question that will give you the most information. When I answer "no," the students quickly realize that they don't really have any information that will help them begin to draw the triangle! As the teachers, we want to get the worse questions asked first, so that all the students to begin to see what type of questions will give them the best information.

The first night's homework is about what kinds of information will be helpful in reproducing the mystery triangle. Students also have to explain what they think is the minimum number of questions needed to correctly guess the triangle. When we discussed it, I thought students would say five—three side lengths, but only two angle measures (because they could figure out the third angle).

Day 2 is the meat of the lesson. On this day, students will play the game in groups of three or four and will come up with their own questions and drawings. We will circulate around the room, answering questions about the triangle and checking students' guesses. Hopefully, by the end of the period, students will have correctly guessed the triangle with only three questions, and will have a sense of what measurements they need.

The homework on the second night is called "Is it enough information?" It presents a series of Mystery Triangle Game scenarios in which a group has answers to a certain number of questions about the mystery triangle, and students have to decide whether it is enough to correctly guess the triangles.

Day 3 is a processing day. We'll go over the homework, and refer back to the experiences that students had on the previous day when they were playing the game. Then we'll take notes on the shortcuts that will guarantee that one triangle is congruent to another, and discuss the ones that don't work (AAA [Angle-Angle-Angle] and SSA [Side-Side-Angle]). The homework for Day 3 is practice recognizing whether two triangles are congruent based on given information.

▌ November 4: Josh's Journal

Betty and I are working really well together. We have spent a lot of time developing curriculum together and know each other's styles very well. Plus, we are completely on the same page when it comes to pedagogy, so we rarely have any disagreements about how topics should be approached. . . . We usually bounce ideas off each other and then improve on them and finalize them. It is very unstructured. We just flow through a topic such as, "what to do on Day 1," moving from generalities to specific questions that should be asked, back to general goals, thinking all the time about how the activities will play out in the classroom. I think these lessons are going to be great.

▌ November 13: Betty's Journal

We're now planning actual materials and worksheets: game-playing worksheets to record questions and drawings, homework assignments, rules of the game, and laminated "mystery triangles." We spent a lot of time thinking up good homework questions that were doable but also thought provoking so that students would begin to process the results of the game independently. We also talked about how we would transition from asking students to "Draw a triangle that matches this mystery triangle" to asking, "Are these two triangles congruent to each other?" These are two different ways of thinking of congruent triangles, and students could get confused. In the game they are drawing one triangle to match a given model. That is different from looking at two triangles on a page with markings on them and deciding if they match.

▌ November 27: Josh's Journal

Well, we finalized everything today. All the copies are made, the triangles are finished, the lesson plans are ready. I am real happy about the outcome. I think this will be great.

Plus I get the chance to see a colleague teach a lesson—which I've rarely done. We've got the other Math I Intensive teachers on board to help with the post-lesson discussion. So, I think all is set to go . . .

December 3: Betty's Journal

Today I taught Day 2 of our lesson, and I watched Josh teach Day 2 as well. A few unexpected things happened. One thing we had not anticipated was how much the students related the game to the topic we just finished, Triangle Relationships. Students were really into asking "What's the length of the longest side?" and "What's the measure of the largest angle?" and then they knew these parts were across from each other in the triangle. They seemed a bit stuck on asking about smallest and largest parts, instead of asking for the measurements of named parts of the triangle.

Another unexpected thing was that the students realized very quickly that you would only need to ask three or four questions in order to draw the mystery triangle correctly. I think the reason that we didn't anticipate this is because we didn't actually play the game out beforehand. What happens is that students will ask the first question, let's say, it's a side length. Then they attempt to draw the triangle. They realize after they draw the one side, that they need to know the angle measures or other side lengths. Then they'll ask for another side length. Now when they try to draw the triangle with two known side lengths, they realize they need the angle in between. Once they have that angle in between, they'll draw the three pieces and realize they just need to connect the ends to make the triangle. It's the process of trying to draw the triangle with limited information that leads them on to another good question that will help them construct the correct triangle.

I really enjoyed observing Josh's class. It was helpful to see how his students did during the lesson, so that I could be more prepared when I had to teach it. I took notes, and tried to come up with questions to ask the class that may help them avoid some pitfalls that some students had in Josh's class. A few students got stuck on asking questions about the triangle's "largest" and "smallest" parts without knowing their names. It was also helpful to me to watch Josh, because he is very businesslike in class. He focused his attention on the students who were working productively, and not on the students who were off task. I think I need to try to do that more in my class. I have one very disruptive student in particular this year who I direct much of my attention to during class. I feel like I'm letting down the rest of the class. But that's another issue. . . .

I'm teaching Day 2 of our lesson in another class on Thursday, and I'm looking forward to it. I am always pleasantly surprised with what students can accomplish if you give them some freedom and responsibility in the classroom. I need to keep reminding myself that students don't need to be spoon-fed all the time.

▌ Discussion Questions

- What were Josh and Betty's goals for their students? What lesson features supported these goals?

- What were Josh and Betty's ideas and hypotheses about how to teach triangle congruence more effectively for their students?

- In looking back, Josh reflected that perhaps they had been too hasty to choose Day 2 for observation. What factors do you consider when choosing which day in a unit will be the research lesson? What would the relative advantages be of using Day 2 or Day 3 as the observed research lesson?

- What did you find useful or interesting about Josh and Betty's lesson development process?

- What do you think Josh and Betty learned about developing research lessons that they might apply or change in their next cycle?

CHAPTER 20 | **How Lesson Study Changed Our Vision of Good Teaching**

Sue Chlebus and Melanie Kellum

Our mathematics department includes teachers of grades 7–12. We are in two different buildings, a junior high school and a senior high school that are physically connected by a joint cafeteria. The only time that we have together as a department is the time allotted for monthly meetings after school and occasional half-days that are set aside for professional development. There has been little time to discuss curriculum, student learning goals, or vertical growth of our curriculum. We viewed our participation in lesson study as a way to address these issues and joined the Lesson Study Communities in Secondary Mathematics project as a way to launch this work. All four junior high teachers and most senior high teachers committed to doing lesson study for the three years of the project. We met as a full team to share ideas, observations, and decide upon the goal for the lesson. We split into middle and high school subteams to work on our research lessons, and so were able to develop two topic-related, age-appropriate lessons in every cycle.

Our First Lesson Study Cycle

The goal of our first cycle was to develop a lesson that allowed the junior and senior high to develop a concept vertically. From departmental discussions about students calculating area and plugging numbers into formulas they memorized but never really understood, we chose to develop a lesson on *understanding* area. Our main question for students was "How do you measure area?" We gave students irregular shapes for the initial activities so they were unable to use formulas to determine the area. We wanted them to understand the idea of a square unit. When we got to the first teaching, we encountered many challenges. The lesson took longer than anticipated. Students were slow to get started. We began with too many problems and did not have time to pull the ideas together at the end of the lesson. But this lesson was important for our team; we had accomplished our first observation and learned from it. This type of exploratory activity was new for many of us, and it became clear that our students were accustomed to having all the information provided and being led through the problem-solving process.

This triggered the idea for our second cycle of lesson study. Instead of concentrating on a particular area of content, we chose to address what we perceived to be a lack of ability of our students to "tackle" a problem. We developed a list of strategies that we thought students could and should use to help solve a problem both at the junior high and senior high level.

Tables and graphs
Algebra
Check and guess
Keep it simple
Lists
Eliminate possibilities
Illustrate
Try it again

Our theory was that a problem with multiple solutions would help students see different approaches to tackling a problem. We tried to choose problems that would generate a variety of solutions both in the process and the type of math used. The handshake problem fit quite nicely. Here is a version of that problem:

Mr. Greeter had 25 students in his first block class. On the first day of class, he asked them to shake hands and introduce themselves to one another. When they were finished, he asked the students, "How many handshakes have just been exchanged?"

This was the first time we ever really brainstormed different ways to solve a particular problem together as teachers. We generated solutions that used charts to help organize the data collected; used technology to plot, analyze data, and find the regression equation; and created algebraic and geometric representations. *It was amazing* to see how many different methods *we* used to solve this one problem. Based on our work, we developed instructional activities that we hoped would generate a variety of student solutions across multiple content areas. We designed the lesson to include only two problems, so there would be plenty of time for solving and for developing the connections among the representations through class presentations and discussion. In the lesson, students came up with the following approaches:

The first person shakes 24 hands, next person shakes 23 new hands, next person 22, and so on

$$24 + 23 + 22 + 21 \ldots + 1 = 300 \text{ handshakes}$$

There are 25 people. Each one shakes 24 hands. 25 × 24 = 600. But this counts every handshake twice, so we need to divide by 2.

$$(25)(24)/2 = 300$$

To graphically illustrate this problem, start with a smaller problem, for example, use five people.

The segments from each vertex represent the handshakes. (See Figure 20–1.)

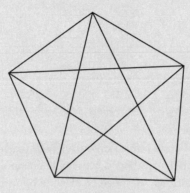

Figure 20–1

Or the number of diagonals plus the number of sides represents the number of handshakes.

$$n\,(n - 3)/2 + n = \text{number of handshakes}$$

Use a table.

Number of person	25	24	23	22	21	. . .
Number of new handshakes	24	23	22	21	20	. . .

Possibly add a column of cumulative number of handshakes.

Several students made this table.

Number of people	1	2	3	4	5	. . .
Total # of handshakes	0	1	3	6	10	. . .

Some students noticed the increasing pattern and continued it.

Other students used a graphing calculator to plot the points on this same table, looked at the graph, and chose a quadratic regression to get the equation $y = .5n^2 - .5n$

Create a chart. X represents a countable handshake for people A, B, C. Continue table.

	A	B	C	D	E
A		X	X	X	X
B			X	X	X
C				X	X
D					X

This lesson had a huge impact on our view of teaching. Instead of expecting one correct process for solving a problem, we should encourage multiple solution methods and use of representations. We need to provide opportunities for students to make connections!

Our Own Learning

Each cycle of lesson study presented an opportunity to expand, evaluate, and improve upon our teaching for student understanding of the concepts and content. It was, however, our fourth cycle that provided an even better experience—that of enriching our individual understanding of mathematics.

We considered probability to be a "weak" link in our curriculum. The junior high curriculum included a unit from *What Do You Expect?*, the Connected Math Project's unit on probability. Other than that particular unit, probability was covered "a little here, a little there." We needed to address the lack of a unified approach to this topic. As we discussed the need for more probability in our curriculum, we came to realize that except for our A.P. statistics teacher, none of us felt confident with our personal knowledge of the subject. So we had a new challenge: to gain an individual level of comfort and competence with the content. Our A.P. stats teacher graciously developed a lesson to teach us, her colleagues.

For a full afternoon, the team worked together at the board, solving problems, using different techniques and debating the outcomes. We examined every type of probability starting with simulations on the graphing calculator. We used the results to determine the relative frequency of the event and estimated the probability. We refreshed our memories on important vocabulary. We studied independent, dependent, and mutually exclusive events. We drew tree diagrams and area models. *We learned together and from each other!*

Brimming with confidence and newly developed knowledge, we were now ready to develop the lessons. Our experience in learning probability greatly influenced the development of our student lessons. We wanted students to experience the "gathering" of data as we had. We wrote problems for both lessons that would require student groups to mimic this routine and ask them to draw conclusions. We led them into discussions of experimental versus theoretical probabilities. We incorporated different problems with different strategies for calculating probabilities so that students would experience playing with tree diagrams, area models, and charts. We spent time working out the solutions to possible problems for the lesson, accepting or discarding problems until we had a diverse sample.

Developing and Observing the Lesson

As it turned out, we were overly ambitious. The lesson had grown into a unit on probability. There was too much to discuss, too many different types of probabilities to ponder, to play with, to diagram. And we wanted our students to experience all of it. Our solution was to create an introductory lesson beginning with simple probabilities building toward more complex problems. We asked our students to play, as we had, with charts and diagrams of the possible outcomes. We hoped that all of this preparation would lead students toward having the skills and knowledge to tackle our one main problem (framed around the Boston Red Sox since they were competing in the World Series, and went on to win for the first time since 1918!). We asked:

What is the probability that the Boston Red Sox will win the World Series?[1]

We were eager to implement this lesson. But as we all know, things don't always go as planned. We asked students to find probabilities for winning the series in 4, 5, 6, or 7 games. Students were overwhelmed and did not know how to begin. We tried to stimulate their thoughts with graphic organizers, but the problem was just too complex. Our expectations were more than what our students were ready for. We tried to include everything *we* had learned in their lesson! In true lesson study fashion we debriefed, revised, and tried again.

---

1   The students were asked to find the probability of the Red Sox winning the World Series, given the following assumptions: that they had already won the division championship, that the winning team was the first to win four games, and that the two teams were equally likely to win any given game.

Lasting Impacts from Our Experience

With all of our individual and combined experiences in teaching, nothing brought us closer together as a department, nor had a greater impact upon improving our teaching than the lesson study experience. Now, we try to anticipate student responses but sometimes students lead us in an unexpected direction. Our work in lesson study gave us the courage to build from student ideas, even if we have to rethink our vision of the lesson, or let go of some of our assumptions. It is not easy for teachers who have been lecturing for years to change the structure of their class. The lesson study experience provided the necessary support and gave us the courage to change.

Our involvement in lesson study across grades 7–12 initiated departmental discussions of the growth of our curriculum. From that, we developed a timeline showing when and what students learn related to a particular content goal. It was surprising to find how much overlap in content we had between the grades, and how much spiraling of the curriculum we really do.

Of course, as a group we had good afternoons when we were "on"—that is, ideas were flowing, with every one of those ideas feeling right and working out well in our planning of the lesson. There were also days when nothing seemed to be right—either we just couldn't come up with an idea that we liked, or we couldn't agree on what we liked. Mathematics teachers (at least in our group) can debate over the tiniest details and sometimes get lost in those details, losing sight of the main goal. We also found that with all of our busy lives, finding a time that worked for everyone without interruptions was difficult. But all in all, our experiences doing lesson study for those three years have continued to impact not only our individual classrooms, but how well we work as a team. The players are stronger and the team is stronger!

Discussion Questions

- How did the team's learning influence their lesson study work together?

- What lessons did the team learn through their initial cycles of lesson study?

- One of the challenges this team faced was focusing the content within a particular lesson. How has your team dealt with this challenge?

CHAPTER 21 | # The Essence of a Day: An Open House Story

Joanne Tankard Smith

NCAS Open House

December 9, 2004

8:15–10:21	Continental Breakfast; Welcome by Superintendent and Principal; Preparation for Observing Lessons
9:36–10:21	First Public Lesson in Mathematics
10:24–11:06	Post-lesson Discussion of the Mathematics Lesson
11:11–12:38	Second Public Lesson in Agricultural Mechanics
12:40–1:45	Lunch and Post-lesson Discussion of the Agricultural Mechanics Lesson
1:45–2:30	Visitation to Other Classes in Plant and Animal Science
2:30–3:00	Closing Remarks by Commentator

Lesson Study Team:

John Lee, Plant Science	Joanne Smith, Mathematics
Melissa McKenna, Animal Science	John Williamson, Mechanics

Team Goal:

Integrate mathematics and agricultural curricula through lesson study.

December 9, 2004

I get these ideas sometimes, like joining the Lesson Study Communities in Secondary Mathematics project and dragging some unsuspecting colleagues along with me. I am never very certain how things will turn out and at some point I might panic a bit. Usually, though, everything turns out OK and everyone benefits in some way. This time is no exception. Today we are about to host our first Lesson Study Open House.

My teammates, our coach, and the staff from Education Development Center (EDC)[1] are here in the cafeteria attending to last-minute details. Earlier I picked up our guest commentator at her hotel and drove her around our campus. Virginia Bastable

---

1 Lesson Study Communities in Secondary Mathematics was a project at Education Development Center, Inc. (EDC) that supported teams in engaging in lesson study in mathematics.

is from Mount Holyoke College and was my mentor at SummerMath for Teachers last summer. I am so delighted that she accepted my invitation to be our outside observer and commentator at this event. I am certain she will provide much insight today. There is a fair amount of energy floating around as we await our open house guests. There is no sign of panic. I watch my three colleagues mingling with the EDC staff and see such confidence in what we are about to do today.

As I think back to August of last year when we were asked if our team might volunteer to be one of the Lesson Study Communities in Secondary Mathematics (LSCSM) teams that would hold an open house this cycle, our team didn't hesitate a moment to say yes. We could see great value in hosting a public lesson. It is an opportunity to put the work we have done together out for scrutiny by our peers. Are we really accomplishing what we set out to do with this lesson? It is incentive to work even harder at our goals. It is a chance to share our school and students with others in a very intimate way. It is a celebration of our efforts and accomplishments through lesson study. The team has worked long and hard to be at this point in time with this lesson and this open house on this day.

I must tell you that we are an unusual lesson study team in this community and have been so from the beginning. When we joined the LSCSM project last school year and attended the introductory workshop, my colleagues found themselves in the middle of a very large room surrounded by secondary math teachers. They were most uncomfortable and cursing my name when we had to do math problems together. I ached for them and admired their courage. You see, I am an academic math teacher at an agricultural high school and my teammates are all vocational teachers. I had been looking for some structured way to integrate mathematics into the agricultural classes and stumbled upon the lesson study work at EDC. I thought this might be just the vehicle for us to get into each other's classrooms and find the math we had in common. I convinced a colleague from each major area of study to join me in the program. So there we were at our first EDC workshop, one math teacher, one plant science teacher, one animal science teacher, and one agricultural mechanics teacher.

Between that first workshop and this Open House we have completed three cycles of lesson study.

> **Lesson Study Cycles Overview:**
>
> Cycle 1—Rates/Stream Flow Estimation lesson and Linear Patterns in Algebra lesson
> Cycle 2—Pythagorean Theorem and Snake Cages—co-teaching a lesson
> **Cycle 3—The Open House: Tiling in wood shop and math class**[2]

---

2 The team conducted a fourth cycle after the Open House, developing two lessons: a lamb-crop percentage lesson and a tractor versus hot-rod lesson on torque.

As a result, the four of us have become so comfortable in each other's classrooms. We have planned, observed, and discussed lessons in each other's disciplines. We really enjoy and benefit from the time we spend together. We are in no way evaluative of each other's work, only of our students' learning. We gain so much insight into our students' thinking in both academic and vocational settings. We meet in pairs or as a group with whatever common time we can find, giving up our prep periods and meeting after school to do this work. We have the full support of our principal who provides coverage for our classes so we can observe a lesson together. We have shared our lesson study work with our school during professional development days and have presented a workshop at the Massachusetts Association of Vocational Administrators annual convention. We are truly unique in the vocational community. Through lesson study we have met our common goal of bridging the divide between ourselves as teachers from different subject areas, different backgrounds, and even different buildings. The trust we have developed is undeniable. We share our worst lessons in hopes of making them better. There is so much less risk when the planning of the lessons is shared and the observation and post-lesson discussion is centered on the students' performance. We feel safe in putting forth today's lesson for public observation. I am so excited to get this open house day started.

Our guests are arriving, teachers from two other teams in the EDC Lesson Study Communities project. And so, our day begins! The food looks good, but I can't eat a thing. My class will be here soon. I hear the superintendent give her welcoming remarks. The principal is introducing the promotional video about our school. My students must be ten times more stressed than I am. I try to watch the video. I keep fingering the lesson handout we made for today. I can't seem to keep my eyes off of the pages. I read the words silently to myself. I don't want to forget anything. Every word, every thought expressed in these pages has been so carefully considered.

It began innocently enough that afternoon when we met to begin the planning for this day. We sat at a worktable, along with our coach, in John's wood shop. I love the wood shop. It smells so good and there is a protractor on just about every piece of equipment! On this day, there was a small stool sitting in the middle of the table. (See Figure 21–1.) At the end of last school year, as we reflected on our first year of lesson study, we decided that the development of measurement sense was a vital topic for all vocational studies and would be a unifying idea for this year's lessons. John had sat the stool on the

Figure 21–1

table because it was his major project for the freshmen exploratory course in the wood shop. He wanted us to help him with the mathematics involved. John and I had worked together exploring the lesson over a few lunch periods, but I was no closer to grasping just what John was after mathematically with this lesson. I hoped that the meeting with the whole team and our coach from EDC would help.

We learned about a lot of things that afternoon. We learned about speed squares, a tool of the trade, and how they relate to protractors. A speed square is a triangle, half a protractor with the curve cut off. (See Figure 21–2.) I was amazed by this tool and thought it might be the first time in a student's life that they encounter a variable scale. We struggled for a long time with a conversation about angle measures. The top of the stool is a rectangle. The stool legs are formed by two pairs of congruent trapezoids that are joined at the nonparallel sides. The pieces are cut from lumber stock. One aspect of the lesson is to cut the pieces with as minimal wastage as possible. Deciding how to lay the pieces out on the strip of lumber is key. As John described the process to us, he kept talking about a 10-degree cut. The rest of us were just not getting it. We went around and around until finally someone asked John to show us in a diagram where the 10-degree angle was. We all seemed to be visual learners at that moment. It turns out that John was talking about a "cut angle," an exterior angle. We were all thinking about the interior angles in the trapezoid! Our angles were actually complements. The moment was profound for us. It simply came down to the meaning of a word. How we laughed! If we are not using language in the same way from shop to math class, we must be really confusing our students. We learned that we needed to speak each other's language first. Our adrenaline went rushing and the conversation began in earnest about complements, supplements, parallel, perpendicular, symmetry, rotations, and translations. It was dark when we finally closed up the shop and left to go home.

Ultimately, we agreed that an exercise in covering related topics at the same time in the math class and the wood shop would be ideal for this lesson and the open house. But I

Figure 21–2

felt no closer to understanding what the part would look like that would play out in my mathematics classroom.[3] I was so relieved when our coach Jane offered to meet with me and work through this together. We met for an afternoon of just doing math together. It was wonderful to be with each other, to talk and to think about the learning involved in the topic at hand. I had forgotten how important this step was for me and how lonely it had been sometimes without another math teacher on the lesson study team. What measurement skills are needed to copy a design onto lumber with accuracy? What thinking processes take place to lay pieces in a linear, end-to-end format that minimizes wastage? As we tried to get our

---

3   The team developed both a mathematics lesson and the wood shop lesson for the open house.

brains around the geometry of tiling quadrilaterals on a strip, a better appreciation of the complexity of the mathematics arose. Recognizing the important mathematics challenged us. We focused on the development of the vocabulary associated with polygons, transformations, and symmetry. We identified ways to experience transformation and symmetry in the plane and to understand tiling. We settled on two questions, "Will all quadrilaterals tile a plane?" and "What quadrilaterals will tile a 6-inch wide strip of lumber?"

The Mathematics Lesson

People are clapping. Someone introduces me and it is my turn to speak about the first of the two lessons everyone will observe today, the math lesson. Somehow, the words come out, the goals, the lesson, the concerns, the hopes. I describe my class, an inclusion class that I teach with a special education teacher. My class usually meets across from this cafeteria in a room small in size and isolated from the main academic building. I am about to bring those students into the cafeteria to sit in groups at tables and to be observed by twenty-plus strangers. I have prepared them, but I'm wondering if moving out of their "home" classroom was a good idea. They will be here soon. I invite the observers to sit at the tables to listen and see up close what is happening in my class. The guests read the lesson handout, ask a few questions, and mill about. I check on the supplies at each table. The bell rings. Here they come.

Everyone has settled at their tables. I have grouped them as heterogeneously as possible hoping they will help each other. I review vocabulary. Good, they are remembering and answering! I hand out the quadrilateral shapes—concave, convex, scalene, isosceles. They will attempt to tile a plane with their shape by copying it multiple times. They will also try to tile a strip, like the lumber they will cut their shapes out of in wood shop later today. They have rulers and protractors to help. I remind them to label their angles with numbers and to remember what they have learned about regular polygons and 360 degrees. Students are attempting to draw, to tile. The measuring is difficult, harder than I thought. The angles on the dart shape are proving to be chal-

lenging. I hope there will be an end product to share! A raised hand, "I don't know what to do next?" I answer, "What questions can you ask?" Work continues. Tiling the strip is interesting. This is novel thinking. Students are posting their results and everyone is walking about to see. Not all students are

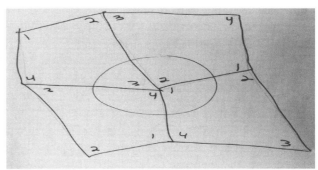

Figure 21–3

convinced that all quadrilaterals will tile a plane. What the characteristics are of quadrilaterals that will tile a strip is also not that clear. The students and I have more work to do tomorrow. The bell rings. They are gone.

Our guests and the team reassemble in another building for the post-lesson discussion. I have the first words and say something about our goals for the lesson. I tell the group that I think the students needed several of each shape to move about, that maybe drawing and measuring was a hindrance. The discussion begins. Someone agrees that the drawing was another level of thinking. Different people noticed different aspects of the lesson. This is great, so many eyes. The students did so many things with the measuring tools and the shapes provided. Observers describe specific students and groups in detail. They recorded bits of conversations among students and described both the accomplishments as well as difficulties faced by the students with the measuring tools. I would never have seen all of this myself. Questions about the mathematics and the learning emerge. How do we build mental images of two-dimensional objects? How do we see shapes fitting together or not? Some students fit several copies of their shape together, put the next one at an odd angle, and declared the shape did not tile! I am thinking more and more that many copies of the shape, shapes that can move, are important for exploring this content.

Virginia assumes her role as commentator. She comments that it is clear this lesson is a step along the way. She says she sees teaching as a constant set of decisions, some made ahead, and some made in the moment. She asks what math each decision about the lesson promoted—what mathematical thinking is involved when students draw verses trace, or if they have one copy versus multiple copies of the shapes to manipulate. Virginia strikes at what I have felt all along, that the mathematics involved in this lesson is much more complicated than we realized at first. Playing with multiple shapes speaks to a different part of the brain, a nonanalytical part. Translating this onto paper is an analytical task. She asks us to think about the responses, "I don't know how" and "It can't be done." These are two different things. How can the structures of the math lesson surface hidden assumptions that are incorrect that we don't know about? This asks us to reexamine every aspect of our lesson in light of what we have heard at this discussion. I am so glad I asked Virginia to come. She knows just how to get at the heart of things pedagogical.

▍The Wood Shop Lesson

I can't believe we have another lesson to do. John briefly discusses the lesson he'll be teaching in the wood shop. Students will have to tile a strip of lumber with the pattern shapes for their stool using the speed square. The bell rings and we all file to the wood shop. I wonder how John is feeling? The first post-lesson discussion went so well and we know this isn't about us as individuals. Some of the students are from my class, but not all. It is a typical vocational class with students from all levels of mathematical abilities. Creating these two parallel lessons has required the team to place much more emphasis on planning together. I feel like I am in John's skin with him as he gives his minilesson and sets his class to work. I watch my students working. They are much more relaxed in the shop, but some still struggle. I see some having success in tiling a piece of lumber with the rectangle and trapezoid pieces of the stool. Transferring pattern pieces onto the wood using the speed square challenges some. Drawing, again. Students make mistakes, make adjustments, learn, and try again. John has wood pieces students can manipulate like a puzzle to check their work. The students are talking easily with each other and helping each other. I don't think this happens with such ease in my classroom. I wonder if my students see the connections between the two lessons? It's that bell again.

It is time to regroup and have some lunch. John gives his thoughts about what happened in his class. Similarities between the two lessons are apparent. Talk centers on having pieces and tools that flip and turn, on the difficulties of measuring with a ruler, and on drawing the figures on the lumber stock. Our observers are surprised just how difficult it is for students to take what is learned in the math classroom and relate it to some real application in another environment. Again, observers have recorded student conversations and remarked on student actions in completing their task. Group dynamics are discussed, particularly about what happens to a group when one student overtakes the ideas or when a good idea is overlooked. Virginia was struck by the importance of visualization to both lessons. What comes first, the visualization or the ability to move pieces around? What happens when you have a piece down, and think that is how it goes, so it is difficult to think about a new way? Where is the stimulation to challenge an idea we have? Where do the new ideas come from? How can we make that an explicit practice for our students within the structures of our lesson?

The discussion is over. Guests are leaving to informally observe a class in animal science or one in plant science with our two other teammates. We take any opportunity to share our unique school and wonderful students with other educators. It will be a tour of the rest of our school and an opportunity for casual conversation and gathering of thoughts. John and I sit together, quietly exhausted. Our guests will be back soon for the end-of-day reflections. I hear them returning. People seem happy. Some have plants in their hands. Others are chatting about holding reptiles. Our teammates join us at the table. The end-of-day reflections begin.

One teacher comments that seeing and talking about lessons is far better than sharing about lessons. I so agree with that. Someone remarks that our students are such risk takers and are very persistent. Somebody says how impressed they are with how hard it is, what we are trying to do for our kids, this integration. Wow! We all understand this today. Virginia remarks that as teachers we don't have to think about the "perfect" thing to do. This is the nature of the work that we do, the ideas and the human beings. It *is* complicated. Virginia closes with how struck she is by how much we want to be in each other's classrooms and how few structures support this. John says that we are required to do integration. Our team has established trust through lesson study. Now that we are deep into the lesson study work we see that our goal is no longer to integrate academics and vocational studies, our goal is to improve teaching.

Epilogue—December 10, 2004

I have given half of the students several copies of identical convex scalene quadrilaterals; and the other half, concave scalene quadrilaterals. They are playing with the shapes individually to see if they will tile a surface. I ask them to find someone with the identical shape and continue to tile. They keep combining with others until there are two tables, one with each kind of shape and most of the surface of the table tiled. We look at the drawings from yesterday's open house that I have tacked up around the room. I relate the numbered angles to 360 degrees and everyone is convinced that all quadrilaterals will tile a plane. Tomorrow we will tackle triangles. The need to practice drawing a figure from measurement remains. My students share their discomfort in being observed, but also describe their pride in being part of the lesson study day. They have a better understanding of how hard we work to create good lessons for them and just how important that is to us.

Reflections

Some surprises for me as I reflected on this Open House included:

- how hard the team worked to prepare the lesson and the day

- how complex the exercise of capturing the essence of the mathematics involved became

- how important it was for me to work with another math teacher

- how easy it was for me to blot out the twenty-plus teachers observing my class

- how uncomfortable it was for my students to be observed

- how much pride in their work the students expressed for having done a good job

- how differently the students see their teachers, invested together in their learning

- how the trust the team has built reached and enveloped our guests

- how focused and contributive the discussions were during the post-lesson discussions

- how our guest commentator deepened our understanding of the lesson

- how emotional the day was for all who participated

Some confirmations for me as I reflected on this Open House included:

- The mathematics of measuring is not trivial.

- The work our team is doing to connect mathematics and vocational studies is important.

- The learning and teaching of mathematics is indeed complex.

- The sharing of it makes for important conversation.

- The lesson is never really finished.

- That we expected the students to "get it" in forty minutes is just hilarious!

How simple it is that observing our students makes us better teachers.

▌ Discussion Questions

- The lessons described in this story include a mathematics lesson and a related wood shop lesson. What was the mathematics in each of the lessons? How were they related? What did teachers learn about the mathematics?

- For each of the lessons described, what do you think the team learned from the observations and post-lesson discussions?

- This story describes the day that the team shares their research lessons in a public open house. What are some of the benefits and challenges of a lesson study open house?

- The team featured is unique in that it is an interdisciplinary team from a vocational high school. What do you think the team learned about the integration of academic and vocational content? In your own work, can you imagine working with teachers in other disciplines focused on integration across disciplines? What could you learn from that?

CHAPTER 22 | The Longmeadow Story: A District Lesson Study Initiative

Euthecia Hancewicz

Longmeadow is a residential community in western Massachusetts whose school district has a reputation for providing high-quality education for its young people. The elementary schools function under an umbrella of shared policies and curricula.

District Initiative

In 2006, Longmeadow's elementary teachers were looking for a way to enhance mathematics teaching in their three grade K–5 elementary schools. They'd had comprehensive multiyear professional development for their English Language Arts (ELA) work and were now looking to focus on mathematics. The district administration and the local parent organization, Longmeadow Education Excellence Foundation (LEEF), had supported the ELA initiative so teachers were hopeful that similar support would make a mathematics initiative possible.

One of the teacher leaders contacted me, knowing that I am an experienced instructional coach who consults with districts about their mathematics professional development. My task was to help figure out a process to work across the whole district. It was quickly apparent that lesson study would be a good match for Longmeadow and its expectations.

Several factors were already in place that would provide a foundation for a district initiative.

- Each school had a mathematics leadership team.

- Each school had a mathematics resource teacher.

- The central administration was eager to provide professional learning opportunities for all teachers.

- One principal had read a lot about lesson study and was an enthusiastic supporter.

- The parent group was able to provide funds outside the regular school budget.

A significant level of funding would be needed for this district implementation. Teachers from three schools were to be involved; money was needed to cover costs for myself as facilitator trainer, substitute teachers, and materials. In many places, these costs must be pulled from the regular school budget. In Longmeadow, the LEEF organization is a private foundation "committed to underwriting grants to provide . . . opportunities for innovative educational and enrichment programs beyond the tax supported budget."[1] A LEEF grant, coupled with funds from the regular school budget, made it possible to move forward with a districtwide plan. Such a combination of school district funds and grants from community organizations is a model worth pursuing in other districts.

Administrative support is always crucial. Here it was a catalyst for success. The superintendent and assistant superintendent created ways for teachers to be released from teaching duties, provided funds for substitute teachers, and encouraged other administrators to participate. They and the school principals attended introductory sessions, some team meetings, and observed many research lessons. This direct interaction allowed administrators to understand lesson study and it honored the work teachers were doing.

One principal's enthusiasm was especially noteworthy. He had read *The Teaching Gap* (Stigler and Hiebert 1999) and was eager to have Longmeadow teachers experience lesson study. Through his relationship with the other principals, the central administration, and the teachers in his building he was a catalyst for this initiative.

Introducing Japanese Lesson Study to the Teachers

The rollout of lesson study in Longmeadow was very different from what often happens. Here we were looking at a whole-district, three-school plan. We were starting with the expectation that all teachers would have the opportunity to be part of a team; within two or three years every teacher would have participated with a team of colleagues under my leadership. The goal was to give teachers enough experience so that they would sustain lesson study as a way of operating throughout the three schools.

The initiative started with one team from each school. All three teams began work in the fall, expecting to complete two lesson study cycles by the end of the school year.

We launched the project with a full-day session for all teams. Teachers learned what lesson study is. They heard about the history of lesson study's growth from Japan to the United States. Many had never heard of it. Several were skeptical. Others were enthusiastic. There were lots of questions and concerns.

Some participants spoke about those concerns, seriously worried that this was just another initiative that would fade away. They were successful teachers confident that their students were learning. They'd seen professional development "fads" come

---

1   Longmeadow Education Excellence Foundation homepage, June 18, 2009, www.longmeadowleef.org/.

and go. They wondered why this apparently simple process could really be helpful—and worth the time!

As the day progressed and participants moved into the actual work of becoming school-based teams, the sounds in the room changed from questioning skepticism to lively engaged conversation. We asked teachers to consider what their students were like and what they would be like in a few years. Teams discussed how to use that information to establish reasonable goals for helping students bridge the gap between current behavior and future behavior. Culminating conversations for the day were about mathematics—what mathematics would be a suitable topic for a research lesson? Teachers were surprised by what the day's discussions revealed about the mathematics and about their work together as a team.

> *I was surprised that the concepts that each grade level wanted to work on were the same.*

> *There are so many ways to look at and discuss one math concept.*

> *I was surprised to see teachers who have taken a back seat at other conferences speak up and be engaged.*

> *Everyone seemed to have an equal voice—all ideas valid.*

Vertical Teams

Longmeadow's intent was to introduce lesson study to the full range of elementary teachers across all three schools, to spread the experience as widely and quickly as possible. We did this through what we called *vertical teams*. During the first year of implementation, each school had one team comprised of a teacher from each grade, K–5, plus the mathematics resource teacher—seven members!

My reaction to the idea was a mix of worry and excitement. Would such large teams be able to come together as one unit? Could I facilitate the shared leadership that's so much a key to lesson study? What a great opportunity for teachers to interact across grade levels! What a great chance for me to learn about the pros and cons of multigrade teams!

The teachers' first reactions to vertical teaming were apprehension; they said things like:

> *I'd get more out of it if I could work with my grade-level colleagues.*

> *It will be challenging to work with so many team members.*

> *How will we ever decide what grade to teach?*

> *As a kindergarten teacher, what can I contribute?*

> *Do I know enough math?*

And there were the unspoken concerns:

I teach the upper grades; I'm not sure what I can learn from the teachers of younger students.

I don't know these teachers very well—and I think that's a problem.

I discovered that large multigrade teams work. I needed to be particularly watchful that each member's voice was heard. Occasionally I had to help a team narrow its focus. One team had to adjust its working norms to ensure participation by all members. All teams figured out how to spread the work across the team. The richness of shared experiences was always energizing.

By the end of the first cycle, teachers were enthralled by the vertical team configuration. Over time, conversations shifted from how unique each grade level's mathematics lessons are to how much connection there is across the K–5 continuum. Participants looked for ways to support student learning through consistent, shared teaching practices. As participants wrote:

Working on vertical teams helps clarify questions and concerns about mathematics teaching.

The process increased feelings of shared accountability for student learning of a concept across grade levels.

One of the great benefits of lesson study was the chance to work on a vertical team of teachers on one concept, and pool our ideas on how to approach the teaching of the concept. To me, that was the best part of the experience.

We came out of isolation and the force of our collective abilities was exciting. Our professional lives were reenergized by the experience.

I am convinced that the vertical team configuration influenced the school climate more than grade-level teams would have done. When teachers know each other well they seek advice from each other. With collegial friendships established beyond "the teacher next door," conversations throughout the school community were richer.

One Team's Story

The seven-person team at Center Elementary School had great potential. An experienced teacher represented each grade level—some were veterans of many years and some were only a couple of years into their teaching careers. The resource room teacher had taught students of many ages and was a respected leader in the school and the district.

During the opening day introduction to lesson study, these teachers questioned the value of lesson study and of the district's commitment to this new initiative. That uneasiness was apparent in team conversations as they worked to set goals and started to plan their lesson. I worried. Would this skepticism hinder the team's work? A couple of team members were exceptionally quiet. I worried. Would I be able to facilitate conversations so their thoughts could be shared and so participants would become adept at listening to everyone's ideas?

The Team's First Cycle

The team's first cycle was a real struggle. The teachers spent a lot of time going around in circles about how to teach a grade 3 lesson on fractions. One teacher offered a favorite lesson as a starting place. Discussions that followed ranged through an amazing labyrinth of mathematical and pedagogical issues. Struggles and discussions continued way beyond what was productive. Finally they settled on a lesson, taught it once, revised, and taught it a second time.

When it was time for the post-lesson discussion following that second teaching, I was worried. It had not been a great experience for anyone. Could I facilitate a productive conversation about the lesson? What would all this difficulty mean for the second cycle and beyond? I was in for a surprise. During the post-lesson discussion of this second teaching, there was such enthusiasm, such a positive response to the experience, and evidence of so much learning that my final reflection on this team's first cycle read,

> The team had a hard time getting started but stuck with the process even when they thought it was going to run them into the ground! The resulting lesson was truly a team lesson with a learning goal that was very important to the group. The range of teaching experience, the variation in philosophy, and the span of grade levels contributed to an extremely rich experience. An amazing amount of work done with a very positive posture!

The Team's Second Cycle

This team's approach to their second cycle was dramatically different from the first. One of the quietest team member's ideas became the basis for the lesson. There were productive conversations about the mathematics and the pedagogy. And there was a deliberate plan to facilitate comfortable contributions from all members.

During the first meeting of cycle 2, team members reconsidered their team norms. They decided to have a deliberate "check-in" a couple of times during each meeting. One member was designated to stop the discussion and ask, "How are you feeling about how this is going?" Most of the time, the answers were, "Fine," but occasionally someone would bring up something that was troubling him or her.

When it came time to decide on the content for the research lesson, one participant told the team what she'd discovered in her analysis of the fourth-grade state test results. Many students were making mistakes on questions about perimeter and area of irregular polygons—specifically on one question in which students were to find the area and perimeter of an irregular H-shaped polygon drawn on a grid.

Rich discussions followed, including input from every team member. I was excited by the range of topics the team explored, which included:

- Standards for grade-level learning about area and perimeter

- How area and perimeter are taught in each grade

- Why there is so much confusion between area and perimeter

- The length of the diagonal of a unit square

- Assessment—formative and evaluative

- The effect of students working individually vs. partner/group work

The team decided to use the problem from the state test in their lesson, but to focus on only perimeter because logistics limited the research lesson to one class period.

We worked individually and collectively to anticipate how students might understand the problem of finding the perimeter of the H-shaped polygon and how they might determine that perimeter. Unlike the discussions during cycle 1, all team members openly participated and all conversation revolved around the problem at hand. Our methods and solutions became the basis of the team's lesson. The first two columns of the research lesson plan (see Table 22–1) show how this played out.

Table 22–1 Research Lesson Plan for Grade 5: Perimeter of Irregular Polygon (Lesson Study in Longmeadow, Center Team, March 2007)

Outline of the Lesson	**What do we expect students to do/ say/think at each step?**
Introduce the lesson with . . . *Main task of lesson is . . .* *Summarize lesson by . . .*	*Anticipated responses* *Students will . . .* *Students might . . .*
Introduction: 5–7 minutes *Vocabulary listed: perimeter, unit, polygon* *• Introduce with an example of sitting at a football or baseball field, around the edge of the ball field is the perimeter.*	*• Something outside of or inside of a shape/ area* *• The distance around a figure*

(continues)

Table 22–1 (*continued*)

Outline of the Lesson	What do we expect students to do/say/think at each step?
Propose the first task: To find the perimeter of an "H"; 15–20 minutes 1. *Ask, "What would you call this shape?" Define it as a* polygon. 2. *Work in pairs using units, crayons, pencils, rulers, or anything else to find the perimeter.* 3. *List all answers; inform students that the correct answer is 26, 22, or 12.* 4. *Say, "Go back with your partner and rework if necessary, be prepared to explain your answer."* 5. *Ask a member from each of the 3 proposed answers to show their work on the overhead.* 6. *Summarize and explain why 26 units is correct.* 	*Polygon, "H," irregular rectangular (an H shape)* • *12—sides of "H" and 12 units can fit inside* • *22—no corners* • *26—each side, the correct answer* • *78—all the background/outside square units*
Propose the 2nd task: 15–20 minutes *Work in partners to solve the question: "What is the perimeter, in units, of this shaded polygon? Describe how you calculated the perimeter."* **Bonus question as the teacher notices the groups that have the correct answer: "Draw and label another shape with the same perimeter."*	*Polygon "Z" irregular rectangular (a right-angle Z shape)* • *8—sides of the polygon* • *20—units to fill the shape* • *22—no corners* • *24—the correct perimeter* • *28—the square units outside the Z* • *24 units, the correct answer: move on to the bonus* • *Bonus answers have various possibilities* • *Some may know the formula and provide it* • *Some may ask: how can it be used for a circle?*
Conclusion: 10 minutes 1. *Go over the correct answer as 24 units, using the overhead to demonstrate.* 2. *If possible, elicit the idea of a formula to determine perimeter of any polygon, s + s + s … = perimeter* 3. *Share any bonus answers and encourage them to try and get a bonus answer on their own.*	

The introductory problem for the lesson asked student partnerships to find the perimeter of an irregular polygon, H, four units across by five units tall, drawn on grid paper. (See Figure 22–1.)

An assortment of manipulative materials was provided—rulers, pencils, centimeter cubes, and blank grid paper. After several minutes, the teacher recorded each partnership's answer on chart paper. Student answers were 13, 26, and 27 units (not the anticipated solutions!). The teacher told the class that the correct perimeter was one of those lengths. She then instructed partnerships to return to their solutions and to prove, in writing, why they were correct. A second problem was presented with the same partnership structure. The lesson ended with consensus about the correct perimeter, sharing of some proofs, and a surprise—students created a rule/formula for calculating perimeter of any polygon.

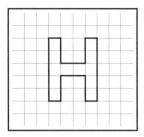

Figure 22–1

Thoughts shared during the post-lesson discussion for this lesson illustrate valuable data collected about student learning:

> *Individual students worked alone and then verified with partners rather than working as a team from the start. If students had been sharing one lab sheet, they would have been more likely to share ideas.*

> *Students were confused when using the ruler because the units didn't match. [The team discovered they'd used two different sizes of grid paper—centimeter and half-inch—without realizing it!]*

> *Using the centimeter cubes was not helpful. Some students figured out the perimeter correctly by inspection, then made errors when they tried using cubes.*

> *Is the use of cubes for measuring linear distance helpful or problematic in relation to students' future understanding of area and volume?*

> *Having students write explanations helped foster the deeper math thinking we wanted.*

This team has continued together through at least six cycles, devoted to the lesson study process and the support they give each other. Their work has included one science lesson and one English Language Arts lesson as well as mathematics lessons. Individuals have also taken on roles beyond their own team. They have invited other teachers to observe their work. They have served as informal mentors for new teams. They see themselves as lesson study leaders.

Conclusion

Each of the start-up teams in Longmeadow has a story to tell. Every team experienced its share of struggles as the teachers worked together creating powerful mathematics lessons and broadening their understanding of what it takes to teach mathematics. Every team became a supportive collegial network across the grades.

Longmeadow's success with its two-and-a-half-year implementation was the result of thoughtful planning and the understanding that effective professional learning takes time. Lessons, such as the following, that we learned here can support other districts as they plan for school or district introduction of lesson study:

- Administrative support is crucial.

- It's possible to get funding from parent/community groups.

- Vertical teams are powerful.

- Teams of six to seven participants are not too large.

Discussion Questions

- This district chose to purposefully include one teacher from each grade level, K–5, on each lesson study team. What advantages did this configuration have? What concerns do you think vertical teams need to address?

- In this chapter, the author notes that "The rollout of lesson study in Longmeadow was very different from what often happens. Here we were looking at a whole-district, three-school plan." How does this implementation of lesson study compare to how you've seen lesson study implemented elsewhere, or to how you can picture implementing it in your district?

- What are the challenges and benefits of implementing lesson study throughout a district all at once? What supports seemed most critical to making it work in this district?

CHAPTER 23 | **Our Lesson Study Journey at King's Highway Elementary**

Anne Nesbitt

▌Improving Students' Standardized Test Scores

In July 2007, the Connecticut Mastery Test (CMT) scores were released by the State Department of Education on a Friday afternoon, at the start of a beautiful summer weekend. The superintendent advised all administrators to look closely at why our district scores did not reflect the capability of our students or the experience of our teachers. As the building math administrator, my responsibility was to focus on assessing the disappointing math results and draft a plan to improve student performance. It felt like the summer sun had temporarily given way to a dark cloud.

I knew that there were several challenges involved in the work that lay ahead. The superintendent's message was to raise test scores, but this did not mean teach to the test. This was not a mandate to embed rigorous test prep into classroom instruction. That was not the culture of our district. For the past four years, we had been committed to exploring ways to improve instruction by focusing on student conceptual understanding. The superintendent's challenge to improve standardized test scores could not compromise this carefully built culture. We needed a plan that:

- *Maintained a math program that supported instruction for conceptual understanding.* The mandate to raise scores could not mean more test prep at the expense of teaching math for deeper student understanding.

- *Provided teacher ownership and time for reflection.* It was important that any message about how to improve test scores be communicated to teachers not as a top-down mandate that usurped their ownership and disregarded their instructional expertise in the classroom.

- *Included analysis of data from multiple perspectives.* It was necessary to unpack the test results by looking for patterns within each concept strand and each objective. We needed to identify the concepts that most grade-level students understood and those that they did not understand, to determine strong and weak objectives within a particular content strand, and to investigate any groups of students that performed outside of the grade-level profile.

Launching the Lesson Study Journey

Serendipity played a big role in launching our lesson study journey as part of our school's plan. I had always been interested in the power of lesson study as a professional development tool. In early 2005 I heard Ana Serrano, a visiting professor at Teacher's College, Columbia University, present a series of lectures on her work with James Hiebert.[1] That fall, she consulted with a group of math teachers at my school who were interested in taking a closer look at the connection between geometric estimation and problem solving using a lesson study approach. In 2006, I was awarded a Japanese Fulbright Memorial Grant to visit Japan. I spent the month of November visiting Japanese schools and asking questions about their lesson study model.

The following April at the 2007 National Council of Supervisors of Mathematics (NCSM) conference in Atlanta, I met Jane Gorman, Education Development Center (EDC) Project Director. Jane was leading a discussion about lesson study and was involved in designing professional development materials for schools interested in starting and supporting lesson study in mathematics. She invited me to participate in a one-week training program during the summer of 2007 for facilitators who would pilot these materials during the 2007–2008 school year.

I attended that facilitator training not long after hearing about the CMT test results and the mandate to make a plan for my school. At the end of the weeklong training, my mind was racing with possibilities. Would I be able to pilot EDC's lesson study model as a professional development tool to increase teacher effectiveness and improve student performance on high-stakes standardized tests? Would I be able to find a group of teachers willing to participate in this research? Would I be able to collect the data to support this implementation plan?

Beginning the First Cycle

During the last week of August, I met with the third-grade team at my school. Third grade was a target grade for improving test scores. We spent some time looking at the puzzling results from the CMT and considered implications for classroom instruction. I described my experience at EDC and invited them to participate in a lesson study pilot. I was convinced that the lesson study model of taking a deeper look at student thinking would provide insight into their students' weak performance on the standardized tests. We shared a spirit of optimism for starting the new academic year using this professional development model. We agreed to a schedule for the fall that included some during-school meetings and some after-school meetings. Recognizing the time

---

1   James Hiebert is one of the authors of Stigler, J., and J. Hiebert. 1999. *The Teaching Gap*. New York, NY. The Free Press.

commitment (approximately forty hours over a semester), the teachers agreed to make this work their Professional Development and Evaluation Plan (PDEP) goal for the year.[2]

The critical task of identifying our lesson study goal was research based. The team referred to detailed spreadsheets created from the 2007 CMT scores to determine what content areas needed the most improvement. There was a range of performance across the five math strands addressed by the test. Nearly 95 percent of the third-grade students performed at the goal or advanced level of understanding for *Numerical and Proportional Reasoning*. However, student results for *Integrated Understandings*, a strand focused on students' problem solving for complex mathematics problems, were markedly different. Only 57 percent of the third-grade students made goal; 43 percent of the children assessed with little or no understanding of how to solve multistep, open-ended, contextually-based problems. It raised a red flag for the teachers. They asked, "How can we provide better instruction to this year's third graders so that they have the conceptual understanding to successfully solve the challenging real-life math problems?" This question became their lesson study goal.[3]

The lesson study process provided a perfect vehicle for investigating this question. The learning cycle included conversations that illuminated gaps in student learning for the group. Through these conversations the teachers gained new insight and subsequently designed a research lesson that encouraged students to think differently. The greatest insight occurred when teachers shifted their perspective from teaching a procedure for solving a particular math problem to teaching the student how to think mathematically—a shift from teaching a math *problem* to teaching a math *student*.

A key experience for the team was to generate a learning map that identified students' prior experience with open-ended problem solving. The team reviewed the second-grade curriculum and made an important observation. Second graders had experience solving problems with multiple solutions, but these problems did not have any limiting conditions. For example, their word problems frequently instructed them to *find as many solutions as possible*, and the students were not asked to satisfy a restriction for choosing among those solutions.

The team considered a third-grade problem that asked students to find three combinations of 5-, 10-, 20-, and 40-pound weights that would each add up to 70 pounds. The complexity of the task was higher than the second-grade tasks because of an additional constraint: the students had to choose exactly five weights that would add up to

---

2   The Professional Development and Evaluation Plan (PDEP) is an agreement between each teacher and the Westport Public Schools. Teachers are required to design and complete a goal for each academic year that demonstrates improvement in student learning. Teachers can work individually or as a member of a team. In this case, the third-grade teachers collaboratively identified the Education Development Center lesson study model as a way to increase student proficiency on state standardized tests.

3   The third-grade teachers cited this lesson study goal in their 2007–2008 PDEP agreement with the district.

70 pounds for each solution.[4] The team collected and analyzed students' written work on this problem. They found that many students were able to meet one constraint of the problem, but not both constraints (e.g., a student might use one 10-pound weight, one 20-pound weight, and one 40-pound weight, which adds to 70 pounds but 5 weights were not used).

Designing the Research Lesson

As the teachers reflected on additional student work samples, it was clear that the majority of third graders did not have strategies for solving more complex problems. The team spent several sessions designing their research lesson to support students' attention to multiple conditions. They realized that this was an area of their instruction that had been missing in past years. The students needed reminders to read the problem carefully, but for some students it was more than an issue of careful reading. These students were aware that there were conditions, but did not know how to revise their thinking. Once they arrived at a solution, and realized that it did not meet a condition in the problem, they could not work backward to correct their answer.

The research lesson was designed around the goal of helping students develop the mathematical flexibility to revise their solution so that it met the problem constraints. The group wanted to use a problem context that was familiar to students and that could be a "hook" into thinking about multiple conditions in a new way. After much discussion, one teacher suggested designing a problem around a book club order. Every third grader had the experience of ordering books through the monthly book club offer. The teachers worked together to create the following problem that mirrored the process of completing a student order form. The teachers created a book order form that showed different books or sets of books for sale, each labeled with a price. Single books were available for $1, $2, $3, $4, $5, $6, and $7. One pair of books was available for $7, two trios of books were available for $8 and $12, and there was one four-book set for $14. The teachers posed the following question to students, and included the four conditions that follow to make it a significant challenge for all students:

> Ms. Nesbitt has decided that our class is going to get $50 to spend on books from the Book Club.
>
> - We must save $5 for a math book that our teacher needs.
> - We need to pay for shipping that costs $6.
> - We need to buy as many different books as we can.
> - We need to spend all of our money.

---

4   This problem was in a test review/practice packet created and distributed by the district math team for assessment and teaching purposes.

The teachers spent considerable time preparing the lesson plan. They wanted to build on student prior understanding. In anticipation of children reading the problem quickly, becoming frustrated, and declaring, "I'm stuck," they included additional questions to prompt student thinking without directing their steps.

- *What are all of the conditions that need to be met?*

- *Can you find one solution that meets all of the conditions?*

- *What conditions does your solution meet? Are there any conditions that it does not meet?*

The teachers paid careful attention to limiting the amount of scaffolding that they planned for this lesson. I had shared with them Jane Gorman's observation from the summer facilitator training, "Student learning is compromised when well-designed lessons are heavily scaffolded." This was an important revelation for the team. They realized that in the past they had sometimes provided too much direction.

The teachers also included questions on their lesson plan to encourage student reflection and strengthen peer discourse.

- *What choices did you make when you were buying the books?*

- *How can you solve the problem in a different way?*

- *What did you learn about problem solving today?*

- *How will this help you be a problem solver/mathematician?*

▌ Debriefing the Research Lesson

The debriefing, or post-lesson discussion, of the research lesson was an invaluable part of the process. Teachers gathered to share their observation notes after the first teaching. They were pleased that the students were so purposely engaged throughout the entire time. They attributed this to the real-world component of the design; the students loved the task of selecting their favorite books. The teachers were also pleased that their anticipation of frustrated students "being stuck" was not an issue. Most of the students were able to understand the problem and recognize that there were multiple conditions. They could at least get started and find one solution.

The more advanced thinkers were able to develop sophisticated strategies. They combined the $5 for a math book and $6 for shipping costs and then subtracted $11 from the $50 pot of money [$50 − $11 = $39]. They then selected as many different books as possible until they reached a total of $39.

The problem was very tricky for those students who ignored these conditions until the very end after they had already spent their $50. They did not make the connection that to have enough money to pay the $11 (for both the math book and the shipping costs) they needed to delete $11 worth of books from their order. The teachers found that their prompting questions provided just enough guidance to have students recheck their work and to recognize their faulty thinking.

The hardest condition for most students was to select as many different books as possible. Many of the students wisely selected books that were packaged together for a bargain price, but only a few students realized that it was more advantageous to spend $5 buying two books (one for $1 and another for $4) rather than to buy just one $5 book.

The teachers revised the lesson to provide more attention to this overlooked condition. They added time to the minilesson to encourage students to think about the concept of trading.

- *How many books can I trade for one $7 book?*

- *How will this information help you solve the problem?*

At the final meeting of the lesson study cycle, although the team was mentally exhausted, they were emotionally excited by the increase in the level of student understanding. One teacher commented for the group, "I know that our students are going to do really well on the CMT *Integrated Understanding* strand this year. They really get it this time! We taught it conceptually; we didn't just teach it procedurally." Lesson study had provided the opportunity to explore the wording of questions and to gather data about where students get stuck during a lesson around a problem with one or more constraints. The team had learned about how their students approached complex problems, and had found strategies that they could integrate into their every day practice to help students tackle these problems.

Reflecting on the Outcomes of Lesson Study Process

In July 2008, when the CMT scores were released by the Connecticut State Department of Education, the superintendent called with congratulations. Our third-grade students made the largest gain of any school in the district! They improved 21 percentage points on *Integrated Understandings*. That good news traveled fast. When we had a chance to meet together in August, before the start of the 2008–2009 academic year, the team was sure that the remarkable success was due in large part to what they had learned about student mathematical thinking during their lesson study experience. The teachers' own words capture best our journey:

The cycle involved choosing a concept, planning a lesson, designing a way to collect data, and then reconvening to evaluate the plan. At each stage in the lesson study, there was professional discourse that helped us refine and revise until we were satisfied with the result.

Lesson study helped me understand the connections to second- and fourth-grade curriculum. I also think it was important to have a chance to plan and observe a common lesson, and then debrief.

I felt that through our discussions I had a better understanding of the range of learners. It made me see that children understand math at different levels. I've never really thought about it that way before. It used to be either they "got it" or they didn't.

It was interesting to see how the lesson we planned together unfolded in the classroom. Many students did exactly as we predicted, while others surprised us.

I learned how important it was to have the time to reflect after a lesson. This reflection allowed us to look more deeply at a concept and thoughtfully consider each aspect of our lesson.

I look back on this lesson study cycle with the same satisfaction that I feel after a long early morning walk. The time spent observing seasonal changes in the flora and catching quick glimpses of the woodland fauna makes the hard work of climbing steep hills so rewarding. This quiet period of reflection stays with me throughout the day, keeping me alert when dealing with the responsibilities of being a school administrator. A lesson study journey brings the same reward. It offers an opportunity to keep student thinking at the core of everything we do, and it gives the hard work of teaching a sense of lasting importance.

Discussion Questions

- How was the team's design for their lesson plan influenced by their discussions and examination of student work? In what ways did this team's revised lesson plan build from their learning about student thinking during the first teaching of the lesson?

- This story starts out by explaining that the district has a focus on supporting conceptual understanding, providing teacher ownership and time for reflection, and analyzing data from multiple perspectives. In what ways was lesson study a good fit for these goals?

- This team saw a dramatic increase in test scores for their students on the strand related to complex problem solving. What about the lesson study process might have helped these teachers change their practice in ways that developed their students' understanding of how to tackle these types of problems?

| # Expanding Lesson Study Practice at Our High School

Debra Casey, Mary Kierstead, Brian Stack, and Brooke Wilmot

Andover High School is a highly academic suburban high school. Most graduates go on to attend a four-year college. The school uses a four-by-four block schedule where most courses are completed in one semester. The lesson study team at Andover High School, including four to six team members at different times, participated in the Lesson Study Communities in Secondary Mathematics (LSCSM) project for three years from 2003–2006. During that time, they completed five cycles of lesson study, hosted two open houses, and attended three open houses at other schools. They also received training to run a lesson study staff development course for other teachers at their high school.

In this chapter, the team relates how they got involved in lesson study, its impact on their teaching and collaboration, and the evolution and growth of lesson study to include other departments in their school.

Starting Lesson Study

At our school, it was the assistant superintendent and the high school mathematics department chair who initially got us involved in lesson study. The department chair offered us the opportunity to join a lesson study team that would participate in the LCSCM project. It was a big commitment—we were signing up for at least two years including after-school meetings and additional days for professional development workshops, and an institute in the summer. Looking back, one of our team members said, "Before I knew anything about lesson study, I couldn't imagine how you could spend two or three months working on one lesson. Now, I have a better sense of that."

To get introduced to the basic process of lesson study, our initial group of six mathematics teachers attended a three-day summer institute in 2003. At this summer institute we saw the video *Can You Lift 100 Kilograms?* that gave us a better idea about what lesson study is. We participated in a number of activities to learn about lesson study and launch our first cycle of lesson study. It was during this workshop that our team began to brainstorm about "what makes a good lesson" and we determined our overarching goal: "To teach students to become active, independent problem solvers and to be open to considering multiple strategies to solve problems." That goal continued to guide our work over the three years we were involved in the LSCSM project.

When we returned to school in the fall of 2003, we began our first lesson study cycle, meeting regularly after school, every other week for two hours throughout the cycle. We continued this basic meeting schedule for the next three years, through five cycles of lesson study. During each of our lesson study cycles a coach from the LSCSM project worked with us. Two different people served this role over the course of the three years, helping us learn about the process and details of lesson study while simultaneously working toward the final product. Our five research lessons spanned the different classes we teach at Andover High School: Algebra II, Geometry, Algebra I, and College Review Math & Problem Solving. Our topics have included: solving systems of equations, a geometry walk, spherical geometry, and problem-solving strategies. A timeline of our lesson study work follows:

Summer 2003	Join LSCSM Project. Attend LSCSM Summer Introduction to Lesson Study Institute.
2003–2004	Complete two cycles of lesson study.
2004–2005	Complete two cycles of lesson study. School offers lesson study course to colleagues in other disciplines. Host first open house—spherical geometry research lesson—spring 2005.
2005–2006	Complete one cycle of lesson study. Host second open house—problem-solving research lesson—winter 2006. Team members offer lesson study course to colleagues.

Attending and Hosting Open Houses[1]

We attended two mini-conferences that were part of the LSCSM project professional development workshops during which lesson study teams shared lessons with each other and we also attended three open houses where we were active participants in the actual observation, data collection, and post-lesson discussion for the host team's

---

1   See Chapter 17, *Public Lessons: The Lesson Study Open House* for more information about open houses.

lesson. These experiences had a tremendous effect on our thinking about our students and our teaching. The open houses inspired us to host our own. We hosted our first lesson study open house in March 2005 and a second in January 2006. The audience actively participated in the process of observing, collecting data, and analyzing the information with us.

Reflecting back on that experience of hosting an open house, one of our team members wrote:

> *This lesson was also special because we ran an Open House for outside educators to come see what lesson study is all about. This professional development opportunity has been so exciting and gratifying that I'm ecstatic to share it with the Andover professional development community and surrounding communities. What could be better than educators discussing student learning and teaching techniques? We are so frequently pressed for time that discussions about education go by the wayside and this open house was a wonderful reminder of what education is all about . . . the students and how to get them to be successful, independent learners.*

▌ Expanding Lesson Study to Other Disciplines

While engaging in our own lesson study work, we also began to spread the word in our department and within the school. We talked with our colleagues about lesson study, we shared about our lesson study work at our school's curriculum council meetings, and several colleagues from other departments attended our first lesson study open house in spring 2005. In addition, in 2005, our assistant superintendent brought in the project leaders from the LSCSM project to lead an in-house professional development course to help teams of teachers in science, English, and history learn about lesson study and develop a research lesson within each of their own disciplines.

In early 2006, at a time when our principal was encouraging schoolwide collaboration, we decided we would seize the opportunity to offer our colleagues an experience with lesson study. Our colleagues had heard about our lesson study work through our presentations and the in-house professional development course that the assistant superintendent had arranged. We decided that the next step was to offer a professional development course to our colleagues, which we called High School Lesson Study. It combined two professional learning community initiatives: Critical Friends Group and Lesson Study. One of our mathematics lesson study team members and a colleague in the history department had been participating together in a leadership class and began talking about similarities between the lesson study work and that of Critical Friends Groups (CFG). Both are aimed at establishing professional learning communities consisting of a group of teachers who come together regularly to improve their practice through collaborative learning. Out of their discussions together came the idea to initiate an in-house professional development course where members of our team would serve as the course teachers and where we would coach other teams as they learned about lesson study and experienced their first cycle developing a research lesson. Initially, there was some tentativeness about bringing together the lesson study and CFG initiatives. We struggled somewhat with fully integrating the two initiatives and ended up focusing more on using the lesson study processes.

Developing Teacher Leadership

To offer the High School Lesson Study course to our peers, we had to "change hats" from being participants in lesson study to being "coaches." We had the full support of our school's professional development committee when we asked to hire our former team coach from EDC to help us learn about and develop our role as coach. One key activity we spent a lot of time on was possible coaching scenarios. We considered hypothetical situations and challenges that might face a coach in a lesson study cycle (e.g., a disagreement about pedagogical approach for the research lesson within a team) and we discussed the possible ways a coach could handle the situation. With the help of our EDC coach, we worked on mapping out a plan for each of the sessions and developed materials for course sessions.

The High School Lesson Study Course

The High School Lesson Study course that we developed was offered to all teachers and we asked that people sign up "as a team within their discipline." Teams from English, Mathematics, Visual Arts, and History signed up to take our course, with a total of about fourteen teachers. Each of us, the original lesson study team members, was assigned as a coach for one particular discipline (English, Math, Visual Arts, History) and worked closely with that team to complete the cycle. We spent part of each course session introducing a piece of the lesson study cycle (for example, setting goals, anticipating student responses, preparing for the observation) to the teachers from all four of these discipline teams, then there was time for the individual teams to work on their lessons. We usually ended each session with some sharing or summarizing, which was very valuable.

Each of us teaching the course took primary responsibility for leading one or two of the course sessions, in addition to our work coaching one of the four discipline-specific teams. Our team member who coached the art team later reflected:

I was the coach for the Visual Arts team of teachers. The fear of working with an unfamiliar discipline made coaching the team intimidating but I was truly excited about the challenge. Collaborating with other teachers is always an inspiring and productive experience, but working with a completely different discipline was over the top. Working with several disciplines set the stage for thoughtful philosophical and pedagogical discussions across the curriculum. To get small groups of teachers conversing about teaching was the greatest outcome of this course. The team of art teachers found the work so rewarding that they asked me to coach them for another cycle. This was humbling.

Working across the disciplines, we learned that we all ultimately have the same goals for our students at some level. The sharing piece was particularly eye-opening because we had wondered how the art team was going to make helpful recommendations for the math team, but they were really able to do that. There was sharing of ideas, and good conversations. Working with teachers in the different disciplines was definitely a highlight of the experience. It was interesting to see the kinds of lessons developed in the different disciplines. We learned something about the structure of other disciplines, how teachers in those disciplines develop techniques and concepts, and what's important in each discipline. We also found it interesting and valuable to see our students in a different setting, especially to see a low-level math student really shine in a different discipline. Seeing another experienced teacher teach was also incredibly valuable, and not something we often have the chance to do.

Benefits of Lesson Study

As a team, we experienced many benefits both from our team's participation in lesson study and through our opportunities to share lesson study with colleagues through the High School Lesson Study course that we developed and led.

The benefits of our participation in lesson study included problems, activities, and ideas that can be applied to teaching in our other classes; greater attention to student thinking; and stronger connections with colleagues. When we reflected on our team's lesson study work, we noted:

I have taken all of the discussions regarding student learning with me into the classroom. I have used three out of the four problems [from our problem-solving lesson] in my ninth-grade algebra class. I have incorporated the problem-solving rubric [that we developed] and the teamwork rubric that we developed with those problems. They have turned into a wonderful learning and team-building activity for the class.

I have grown as an educator and feel very lucky to have been involved with lesson study from the get-go. I've been teaching just as long as I've been a member of this initiative and am all the better for it. I am grateful for the chance to work with more experienced teachers and to have the time to think about what is important to me as an educator.

My participation in Japanese Lesson Study has had a long-term and significant effect on my professional career and on the students I teach. Lesson study is a powerful tool for teachers, emphasizing collaboration with peers, improving instruction, and developing methods to encourage active thinking, all of which have a dramatic impact on student learning. It has been at the core of my professional development and growth as a teacher of mathematics thus far.

While I worked with my peers to complete five cycles of lesson study, I took the opportunity to look more deeply into topics I wouldn't have previously made the time for, strengthening my knowledge and depth of understanding of mathematics. Working with others in this productive environment gives one a new perspective into the content area, as well as into the art of teaching. Lesson study is truly a spectacular environment to mold one's practices and to gain confidence. It has encouraged me to explore the opinions of colleagues, regardless of the discipline, and to truly listen to their perspective when it comes to how students learn best.

Our collaboration as a lesson study team has opened a lot of doors. Personally, I am not a strong mathematics person but after interacting with the other team members in our lesson study group, I now feel comfortable going to any of my colleagues. I never feel intimidated to ask [a veteran teacher] questions. The lesson study setting made it comfortable. We became a real team.

Furthermore, leading the High School Lesson Study course was a valuable learning experience for us. Reflecting back on the experience of leading the course and coaching another team, one of our team members wrote:

[Leading] the High School Lesson Study course was empowering and rewarding. I find it inspiring to work with other teachers and get ideas from them, which translates into my enthusiasm in the classroom. I discover new activities and ideas that I can weave into my curriculum and daily classroom activities that change the effectiveness of what I am teaching in the classroom. It was rewarding to offer this professional development course for my colleagues knowing the positive effect it has both professionally and personally for participants.

One of the greatest contributions and effects this experience had at the time was on my students. During the year I was coaching, I participated in six observations and it was during these observations that I observed strengths and weaknesses of students—both personal and academic—that I hadn't observed in the mathematics classroom. I transferred these observations into the mathematics classroom, talking with students about their accomplishments in art or encouraging a student to work with another person if I noticed he/she was a leader in the art classroom.

The experience also gave us a greater appreciation for our colleagues at the high school. We found them to be insightful and interesting people who have deep knowledge of their subject matter and also have the drive to continue to learn. Getting to know and appreciate these colleagues was an exciting part of the work.

Administrative Support

Over the several years we participated in lesson study, our team has enjoyed the support of district and school administrators, especially the support of the district assistant superintendent and the department head for mathematics at Andover High School. We firmly believe that the assistant superintendent was a key to the success—having that support at such a high level in the system allowed our team work to thrive and opened up opportunities for us to grow as leaders while sharing lesson study opportunities with our colleagues.

There was also another important systemic support for our work. While we meet after school, we were able to receive district professional development course credit for the hours that we spent on our lesson study work. The district had an established structure for in-district professional development courses led by teachers. Andover teachers earn credits by participating in and leading professional development courses. These credits count toward track changes that count toward salary increases. So teachers were highly motivated to participate in quality professional development. Obviously, the course was also hard work, and attracted a group of people who wanted to have a good learning experience—not just easy credits—but the fact that the district provided that support at all was incredible.

What's Happening Now?

In the past couple of years some shifts have affected our lesson study work. The assistant superintendent has left the district, a couple of the teachers in our group left the school, the LSCSM project that we were involved with ended, and our lives and commitments (e.g., the courses we are teaching) have changed. Without some of the original supports that were helping to make our work possible, it has been hard to sustain the use of lesson study in our school. However, taking the time to reflect on our experience over those five cycles of lesson study and during our work as coaches supporting other teams, we are remotivated, and the question of how to get something going again is certainly on our minds.

Discussion Questions

- The team describes many benefits that they gained from their lesson study experiences. What impacts seemed to be the most powerful? What benefits have you experienced or do you think you could gain from your lesson study work?

- The team's leadership within their school grew as they developed their own lesson study practice and began sharing about their work with others. How did the team exercise their growing leadership? What helped the team develop their leadership capacity—as a team and as individuals? What opportunities do you see for teacher leadership in lesson study practice?

- One of the motivations for the team to share their learning about lesson study was the opportunity to connect with their colleagues in other disciplines. What did the team learn through their cross-disciplinary work?

| **Our First Open House: Exploring Spherical Geometry**

Debra Casey, Mary Kierstead, Brian Stack, and Brooke Wilmot

In the 2004–2005 school year, our lesson study team included four mathematics teachers and we were in our second year participating in the Lesson Study Communities in Secondary Mathematics (LSCSM) project. The previous year we had two additional team members. We made a decision not to ask new teachers to join our group to replace the teachers who left because we wanted to build on what we had already learned and continue to develop our own lesson study work. We were excited to share what our team had learned together through lesson study by organizing a lesson study open house.

In our spring 2005 lesson study cycle, we worked on a geometry lesson that would be featured at our first open house. Our school, Andover High School, is an academic-focused suburban high school. Most graduates go on to attend a four-year college. The traditional sequence of courses is Geometry in the freshman year, Algebra II and III in the sophomore year, and Precalculus in the junior year. Seniors are offered the opportunity to take courses ranging from Calculus and Statistics to College Review Math and Problem Solving. All courses are leveled as follows: enriched, level one, level two, and level three.

This research lesson was designed for both the enriched and level one geometry courses, and corresponds to part of a unit in our textbook on exploring proof (Bass et al. 2001). Topics in this unit include deductive reasoning, types of proofs, parallel line theorems, and spherical geometry. The lesson was designed to be completed in an eighty-two–minute block.

Administrative Support

We have had strong support from the administration (including our math department head, the principal, and assistant superintendent), other teachers in the school, and from the Education Development Center, Inc. (EDC) staff of the LSCSM project.

Choosing Goals and Topics

Our long-term goal is to teach students to become active, independent problem solvers and to be open to considering multiple strategies to solve problems. We have developed our lessons with this in mind. During our first cycle, we chose the topic of solving a system of linear equations by graphing, in the context of word problems. Last spring, we undertook an ambitious two- to three-day lesson that included a geometry field trip during which students used geometry to answer a collection of real-world problems. Last fall we prepared an Algebra I lesson, the focus of which was following directions and cooperative learning.

The lesson we prepared for our first open house in spring 2005 was a lesson in spherical geometry. It was an activity-based lesson in which students explore a new topic related to, but different from, the more familiar planar geometry. This lesson had three major objectives:

- Students will understand key terms from spherical geometry.

- Students will understand some of the key similarities and differences between Euclidean and spherical geometry, including the triangle angle-sum rule, similarities and differences of parallel lines in the two systems, and so on.

- Students will gain a better appreciation for such Euclidean postulates as the Triangle Inequality Theorem, the Parallel Postulate, and the Triangle Angle-Sum Theorem.

Developing the Lesson

We consulted a variety of different resources including our own textbook to see how it handled spherical geometry. We spent time at our meetings working through the spherical geometry problems from three different textbooks. We found one of the key activities in our lesson, a hands-on activity investigating parallel lines on spheres using a beach ball, in one of the textbook resources we reviewed. Another benefit of doing the mathematics problems together was that it was a refresher on the mathematics for us and strengthened our own understanding of non-Euclidean geometries, and spherical geometry in particular. Working on the mathematics problems that we considered for the lesson was challenging and several times we needed each other's help to solve the problems. We had fun doing the math together!

A big part of our conversations centered around the purpose of exploring spherical geometry in a geometry course. We wanted our students to learn that there are geometries other than the Euclidean geometry that we'd been studying together all year. We thought that a good way to expand students' thinking was to compare the assumptions that are different in these two geometries, and examine the different rules or theorems that arise as a result. To help students gain a better understanding of how to think about geometry on a sphere, we wanted them to consider questions such as: What is a line on a sphere? What are parallel lines? What is the sum of the angles in a triangle?

In addition, we wanted students to understand that spherical geometry is just one type of non-Euclidean geometry, not the only type.

We designed a lesson that was a mixture of lecture, discussion, cooperative learning, and investigation, with the following organization:

- Students will complete a homework assignment prior to the lesson that introduces some of the spherical geometry terms that will be necessary in the lesson.

- Following discussion of the homework assignment, students will explore the connection between parallel lines and spheres through an activity.

- The activity will lead to a discussion about the major similarities and differences between Euclidean and spherical geometry. Students will investigate these similarities and differences using basketballs and string. Information will be recorded on a chart. Further exploration and example problems will be reviewed.

- Finally, students will complete selected problems for homework.

Attending Lesson Study Open Houses

We attended an open house at Norfolk County Agricultural High School in the fall and observed two research lessons: a geometry lesson focused on tessellations (tiling the plane with quadrilaterals) and a related lesson held in the wood shop class that dealt with laying out the pieces of a project on wood to minimize the number of cuts and material waste. (See Chapter 21, *The Essence of a Day*.) This was the first time that we had participated in an observation and post-lesson discussion that involved a much larger group. We got to hear many different perspectives on the lesson observed and we saw how useful those points of view could be in revising and improving the lesson. Participating in the open house energized us to want to offer our own, and observing the math and carpentry lesson spurred our interest in cross-disciplinary work.

We also attended an open house at Watertown High School in March. There we were really excited to see the quality of the mathematics lesson that they developed. The research lesson focused on combinatorics and the relationships between combinations and Pascal's triangle. An intriguing aspect of that open house was the mathematics session presented by a university mathematician, which extended the mathematical ideas in the lesson that was observed. The speaker inspired us to invite a mathematician known for his work in problem solving to our team's second open house in January 2006. Since the Watertown High School open house we have distributed the research lesson that their team developed to other algebra teachers at Andover High School for their own use.

The Open House Day

For all of the teams in the LSCSM project, hosting open houses was a novel experience—a unique opportunity to share in a larger forum the work we had been doing together to improve our teaching and our students' learning of mathematics. We put our ideas and lesson plans forward for public scrutiny and for shared learning. The structure of our open house follows:

Agenda—March 23, 2005

7:30	Breakfast in the library.
8:00	Opening remarks and an introduction to lesson study.
	Marcia O'Neil, Assistant Superintendent, Andover Public Schools
	Peter Anderson, Principal, Andover High School
8:15	An introduction to our lesson: Exploring spherical geometry. We will take a look at some of the concepts that will be used in our lesson, and discuss what we will need to look for in the observation of the lesson.
9:15	The lesson. Deb Casey, AHS team member, will teach the spherical geometry lesson with her enriched geometry class in the library. Open house participants will be invited to observe this lesson.
10:45	Break.
11:00	The open house participants will be given the opportunity to meet as a group to discuss the lesson presented in Deb's class.
12:00	Lunch will be provided in the library.
12:45	Block 4. Open house participants are invited to observe a regular classroom lesson in one of the following classrooms:

Biology	Steve Sanborn	Room 217
Dramatic Literature	Eric Pellerin	Room 201
Integrated Math I	Brooke Congdon	Room 343
AP Statistics	Mary Kierstead	Room 359

2:15	Wrap-up and closing remarks in the library.
3:00	Adjourn.

The opportunity to share learning is exemplified by the post-lesson discussion that we had with our team and our invited guests. Everyone involved had collected rich data about how the students were thinking about spherical geometry. Examples of what we learned by examining those observations included that we might have included too many problems in the lesson because we were trying to do so much—we needed to be clearer about our goals for student understanding for that one lesson and narrow our focus. In addition, our observations helped us to see how important the hands-on experience with basketballs and string was for the development of student under-standing of parallel lines in spherical geometry.

Reflections on the Day

One way in which the open house changed our work is through the variety of per-spectives that were represented. We invited other mathematics, science, and English teachers from Andover, administrators, and mathematics teachers and administra-tors from outside of Andover. We have been taking leadership in helping others learn about lesson study in Andover,[1] and invited other school systems to the open house to increase their understanding of lesson study. This diverse group of participants in the day impacted the discussion and produced a wide range of observations about the lesson. We received valuable feedback from these observers, and their presence increased our learning from the lesson. It was particularly interesting to hear the ideas for revising the lesson from the teachers of other disciplines. Their reflections about our learning follow.

On Being the Teacher

This year, I took on the role of the "teacher". . . when people hear about how lesson study works, they always think it would be hard to be the teacher, with observers watching your every move. Over the years I've come to realize that the observation has nothing to do with the teacher and everything to do with the lesson. I've also dis-covered that the more observers there are, the less I notice them as a teacher. At our open house, we had some thirty observers, and I don't think I noticed any of them. The general reaction from my students was that after about ten minutes or so, they didn't really notice them.

On the Open House Post-lesson Discussion

Our lesson study open house was the culmination of not just one cycle of work, but of our years of experience with lesson study. . . . I was proud of the lesson we created. . . . [The visitors] were excellent observers and had very thoughtful comments from their own teaching experience to share. I learned quite a bit from their feedback and from the [post-lesson discussion]. Every time I have gone to an open house at another school, I have left the conference feeling inspired and excited about teaching. I felt that way after this open house as well, even though it was our own.

---

1   See Chapter 24, *Expanding Lesson Study Practice at Our High School.*

On Sharing the Work

This lesson was also special because we ran an open house for outside educators to come see what lesson study is all about. This professional development opportunity has been so exciting and gratifying that I'm ecstatic to share it with the Andover professional development community and surrounding communities. What could be better than educators discussing student learning and teaching techniques? We are so frequently pressed for time that discussions about education go by the wayside and this open house was a wonderful reminder of what education is all about . . . the students and how to get them to be successful, independent learners.

On the Structure of the Open House

Our team spent some time reflecting on the structure of our open house, and certain aspects of our plan for the day seemed to have contributed to its success. We allotted time for the open house participants to study the problems before they were presented in the lesson. This way, the participants would get a firsthand experience of what the students were going to go through during the lesson. We also assigned observers to specific roles during the lesson (i.e., to observe specific groups or to roam, and to look for data regarding specific questions). This made the observation much more effective because it gave each observer a chance to study a particular item in more depth.

On Teacher Learning[2]

The teachers here are calculated risk takers . . . willing to step forward if something looks like a good learning opportunity. We ask students to take risks but as adults aren't always willing to. [Offering a lesson study open house] was a risk, but one that yielded lots in return. Here at the high school, ability, excitement, passion has started to multiply. These teachers are teaching their colleagues.

Following the Open House

Having had a valuable learning experience hosting our first open house, our team went on to host another in January 2006. We were excited to begin the process of expanding lesson study at our school. We ran a staff development course at our school, and each of us played the role of a lesson study coach for four new lesson study teams, all in different disciplines. In the following year, the school district asked that we open up

---

2   These thoughts were shared by the district's assistant superintendent at the team's open house.

the course to both high school and middle school teachers. We'd like to see lesson study expand to all grade levels, and influence the nature of professional learning communities at our school.

Discussion Questions

- How did this team's goals drive the team's lesson development, observation, and discussion?

- What did you learn about how the team developed its research lesson? How is that similar or different from what your team has done?

- What did the team gain from attending open houses at other schools? What do you think you could learn attending an open house and observing and discussing a research lesson at another school? How would teaching change if teachers had regular opportunities to observe teaching practice at lesson study open houses?

- What were the benefits of hosting an open house for the team members? How did offering an open house affect their work? What might have been the benefits for participants attending the open house?

Resources Appendix

Lesson Study Resources

- Lesson Study Websites

- Understanding Lesson Study—Readings

  - Brief Readings for Teams and Leaders

  - Handbooks and Facilitator Guides

  - Readings on Lesson Study and Mathematics Teaching in Japan

- Lesson Study Videos

  - Video Introductions to Lesson Study

  - Video and Lesson Plans of Research Lessons in Mathematics

Lesson Study Websites

- *www.edc.org/lessonstudy*   This is the website for the EDC Lesson Study Center[1] and contains general resources for lesson study, information and resources about the Lesson Study Communities in Secondary Mathematics project, and local lesson study groups and workshops.

- *www.lessonresearch.net*   This website, from the Mills College Lesson Study Group, under the leadership of Catherine Lewis and Rebecca Perry, contains a wide array of lesson study resources, including a number of excellent lesson study videos. Two of these (*Can You Lift 100 Kilograms?* and *How Many Seats?*) are introductions to lesson study, and can be used as alternate videos in *A Mathematics Leader's Guide to Lesson Study in Practice*, Session 1. The website also links to national lesson study sites, workshop resources, and an extensive set of excellent research articles on lesson study.

- *http://hrd.apecwiki.org/index.php/Classroom_Innovations_through_Lesson_Study*   This website, managed by the Asia Pacific Economic Community

---

1   Learning and Teaching Division, Education Development Center, Inc., 55 Chapel Street, Newton, MA 02458.

(APEC) Human Resource Development Working Group, contains a glossary of lesson study terms, videos of research lessons from many countries (with subtitles), and articles about lesson study and mathematics teaching. It also provides U.S. teachers with a window into the international mathematics education and lesson study community.

- *www.lessonstudygroup.net*   This website is from the Chicago Lesson Study Group at DePaul University, under the leadership of Dr. Akihiko Takahashi, and contains information about the group's annual lesson study conference, as well as many research lessons and resources from the group. This group supports a lesson study listserv through which practitioners can share questions and information.

- *www.globaledresources.com*   Global Education Resources, under the leadership of Dr. Makoto Yoshida, provides a list of textbooks and teacher guides from Japan (translated into English) as well as resources on lesson study.

- *www.tc.columbia.edu/lessonstudy/index.html*   Although no longer actively maintained, this website remains available and contains a large store of lesson study resources and research developed by the Lesson Study Research Group at Teachers College, Columbia University.

- *www.uwlax.edu/sotl/lsp*   Lesson Study Project from the University of Wisconsin-LaCrosse hosts a site that is devoted to lesson study for college faculty and includes basic resources, reports from teams, and interviews.

Understanding Lesson Study—Readings

Brief Readings for Teams and Leaders

Lewis, C. 2002. "What Are the Essential Elements of Lesson Study?" *The California Science Project Connection* 2(6): 1–4. An excellent short article that explains the essential features of lesson study including, for example: deepen our subject matter knowledge, think deeply about our long-term goals for students, develop powerful instructional knowledge.

Lewis, C., R. Perry, and J. Hurd. 2004. "A Deeper Look at Lesson Study." *Educational Leadership* (February): 18–22. This article discusses the pathways that connect lesson study to improvement in instruction. For example, lesson study gives teachers a stronger knowledge of their curriculum, which translates to better teaching in general.

Lewis, C., R. Perry, I. Hurd, and P. O'Connell. 2006. "Lesson Study Comes of Age in North America." *Phi Delta Kappan* 88(4): 273–81. The authors describe the history and impact of sustained lesson study in one California district, drawing insights that apply to any district considering lesson study implementation.

Lewis, C., and I. Tsuchida. 1998. "A Lesson Is Like a Swiftly Flowing River: Research Lessons and the Improvement of Japanese Education." *American Educator* (Winter): 14–17, 50–52. This article was one of the early publications that introduced lesson study to educators in the United States. It offers an engaging introduction to lesson study process, philosophy, and Japanese origins.

Takahashi, A., and M. Yoshida. 2004. "Ideas for Establishing Lesson-Study Communities." *Teaching Children Mathematics* 10(9): 436–43. This article provides a clear overview of lesson study with clear diagrams and photographs to help the reader visualize the process.

Research for Better Schools. 2002. "What Is Lesson Study?" *RBS Currents* 5(20). This issue from Research for Better Schools contains a whole series of brief articles explaining the nature and potential of lesson study. Voices of teachers and principals are included.

Handbooks and Facilitator Guides

Gorman, J., J. Mark, and J. Nikula. 2010. *Lesson Study in Practice: A Mathematics Staff Development Course.* Portsmouth, NH: Heinemann. Activities, materials, and facilitator instructions for a lesson study course that embeds a guided cycle of lesson study in mathematics are included. Developed in conjunction with *A Mathematics Leader's Guide to Lesson Study in Practice.*

Lewis, C. 2002. *Lesson Study: A Handbook of Teacher-Led Instructional Change.* Philadelphia, PA: Research for Better Schools.. This handbook provides an overview of lesson study, its origins and philosophy, and well as practical guidance and resources for teams and leaders to start lesson study.

Stepanek, J., G. Appel, M. Leong, M. Turner Mangan, and M. Mitchell. 2007. *Leading Lesson Study: A Practical Guide for Teachers and Facilitators.* Thousand Oaks, CA: Corwin Press, Northwest Regional Educational Laboratory, Learning Point Associates.

Wang-Iverson, P., and M. Yoshida, eds. 2005. *Building Our Understanding of Lesson Study.* Philadelphia, PA: Research for Better Schools. Essays to deepen understanding of specific aspects of lesson study, such as the role of the "knowledgeable other."

Wiburg, K., M. Wiburg, and S. Brown. 2006. *Lesson Study Communities: Increasing Achievement with Diverse Students.* Thousand Oaks, CA: Sage Publications.

Readings on Lesson Study and Mathematics Teaching in Japan

Lesson study teams and leaders have drawn inspiration and knowledge about the lesson study process by reading about its practice in Japan. Lesson study originated in Japan, and has been used for decades to explore teaching through problem solving.

Fernandez, C., and M. Yoshida. 2004. *Lesson Study: A Japanese Approach to Improving Mathematics Teaching and Learning.* Mahwah, NJ: Lawrence Erlbaum Associates. This book provides a detailed description of one entire lesson study cycle in a Japanese elementary school and gives insights into Japanese lesson study practice.

Isoda, M., T. Miyakawa, M. Stephens, and Y. Ohara, eds. 2007. *Japanese Lesson Study in Mathematics: Its Impact, Diversity and Potential for Educational Improvement.* Hackensack, NJ: World Scientific Publishing Company. Brief essays describing aspects of lesson study practice in Japan and internationally.

Lewis, C., and I. Tsuchida. 1997. "Planned Educational Change in Japan: The Shift to Student-Centered Elementary Science." *Journal of Education Policy* 12(5): 313–31. This article describes how lesson study, when used widely and systematically, impacted science instruction in Japan.

Shimizu, Y. 1999. "Aspects of Mathematics Teacher Education in Japan." *Journal of Mathematics Teacher Education* 2: 107–16. Through description of one Japanese mathematics lesson on multiplication, the author summarizes a common Japanese lesson structure and the role of the teacher in each part of the lesson.

Stigler, J., and J. Hiebert. 1999. *The Teaching Gap.* New York: The Free Press. The authors summarize and analyze findings from the Third International Mathematics and Science Video Study comparing teaching in Japan, Germany, and the United States. This book makes a powerful argument that lesson study has potential to change the culture of teaching, improve instruction, and build a professional knowledge base. It sparked widespread interest in lesson study in the United States.

Takahasi, A. "Characteristics of Japanese Mathematics Lessons." This article offers an overview of a problem-solving lesson model common in Japan. (Available at www.globaledresources.com/resources.html.)

▌ Lesson Study Videos

Video Introductions to Lesson Study

Can You Lift 100 Kilograms?[2] A video introduction to lesson study featuring a Japanese team teaching a lesson on levers in a fifth-grade classroom in Tokyo (dubbed in

---

2   Recommended as alternate introductory video for *Lesson Study in Practice: A Mathematics Staff Development Course* by Jane Gorman, June Mark, and Johannah Nikula (Portsmouth, NH: Heinemann, 2010).

English). Video shows team meeting, lesson, and post-lesson discussion. (Available at www.lessonresearch.net.)

Introduction of Lesson Study. (Akihiko Takahashi) This introduction is presented through slides with voice narration by Heather Brown. (Available at http://hrd .apecwiki.org/index.php/Video_Introduction_of_Lesson_Study.)

Lesson Study: An Introduction. (Yoshida and Fernandez) This video gives a general descriptive overview of the lesson study process with video footage from Tsuta Elementary School, Hiroshima, Japan. (Available at www.globaledresources.com.)

Mills College Lesson Study Group. *How Many Seats?: Excerpts from a Lesson Study Cycle.*[3] A video introduction to lesson study featuring a fourth-grade research lesson on patterns in San Mateo, CA, showing all parts of the cycle: team meetings, two teachings, and post-lesson discussion. (Available at www.lessonresearch.net.)

Mills College Lesson Study Group. *Three Perspectives on Lesson Study.* Three lesson study researchers (Catherine Lewis, Clea Fernandez, and James Stigler) answer a series of frequently asked questions about lesson study. (Available at www.lessonresearch.net.)

A Users Guide to Japanese Lesson Study: Ideas for Improving Mathematics Teaching. Curcio, F. R. (2002). Reston, VA: NCTM Brief user's guide booklet and seven-minute video of Japanese public research lesson and post-lesson colloquium.

Video and Lesson Plans of Research Lessons in Mathematics

The Asia Pacific Economic Community Initiative Classroom Innovations Through Lesson Study has posted classroom video and lesson plans for elementary mathematics lessons from many countries, including several taught by teachers in the Chicago Lesson Study Group. (Available at http://hrd.apecwiki.org/index.php/ Classroom_videos_from_Lesson_Study.)

Research lesson videos (with lesson plans) in elementary mathematics and science are published by the Mills College Lesson Study Group. (Available at www.lessonresearch.net.)

Shimada, S., and J. Becker, J. (eds.). 1997. *The Open-Ended Approach: A New Proposal for Teaching Mathematics.* Reston, VA: National Council of Teachers of Mathematics. This book presents a vision of instruction using open-ended problems and a

---

3   Recommended as alternate introductory video for *Lesson Study in Practice: A Mathematics Staff Development Course* by Jane Gorman, June Mark, and Johannah Nikula (Portsmouth, NH: Heinemann, 2010).

collection of research lessons in elementary and secondary mathematics developed to study that pedagogical model. For each lesson, the reader sees a detailed plan, results of the instruction, and teachers' research conclusions about the open-ended approach.

Teaching Mathematics in Seven Countries: TIMSS Video Study. The TIMSS research included videotaped mathematics and science lessons from eighth-grade class-rooms in Australia, the Czech Republic, Hong Kong, Japan, the Netherlands, Switzerland, and the United States. These lessons were not research lessons, but provide an excellent resource for lesson study teams. Published and distributed by Research for Better Schools. (Available at www.rbs.org/Special-Topics /International-Studies-in-Mathematics-and-Science/44/.)

Many lesson study projects that maintain websites post research lesson plans online. See lesson study URLs for additional research lesson videos and plans.

Mathematics Resources

- *Standards:* Teams begin a cycle by learning about standards for their topic.

- *Textbooks:* Teams study various textbook approaches to teaching the topic.

- *Problems and Lessons:* Teams seek good problems for their own learning and to use in lessons. Similarly, teams look for the best available existing lessons as a starting point for their own research lesson.

- *Research Themes:* Teams may stay with one major research theme, like *algebraic thinking*, for many cycles and seek resources on that theme.

- *Readings from Research:* Teams build their lessons based on current research and seek to contribute their findings to the professional knowledge base.

Standards

Each state has standards documents that lesson study teams consult. In addition, professional organizations have produced standards that are intended to guide teachers from all states.

National Council of Teachers of Mathematics (NCTM), www.nctm.org. Publications and online resources from NCTM set the goals teachers are seeking to achieve for their students and their own learning.

- *NCTM Principles and Standards for Teaching and Learning* provides learning standards, guidance on typical learning trajectories, understanding goals, and sample problems that challenge student thinking.

- *NCTM Curriculum Focal Points for Pre-kindergarten Through Grade 8 Mathematics: A Quest for Coherence* gives guidance in prioritizing among the hundreds of standards teachers may be accountable for teaching. The key topic focus points for each grade are explained, with examples showing mathematical connections.

- NCTM *Focus in High School Mathematics: Reasoning and Sense Making*. This document argues for a focus in high school on reasoning and sense making, and provides examples of how this might look.

- *Achieve.* The *Mathematics Benchmarks K–12* (available at www.achieve.org) provides an alternate set of national standards that specify the core knowledge students will need at each grade level to ultimately be ready for college and the workforce.

Textbooks

Textbooks and the teachers' guides that accompany them are a primary resource for teams. Teams use them to learn about the mathematics, to analyze how topic understanding develops over time, and to search for excellent problems and lesson plan ideas. Teams study textbooks from their own grade level, as well as those mentioned here.

Some curricula are designed around a problem-solving model that many teams use in their research lessons, and so have served as rich sources of problems and lessons for teams. A few examples of curriculum available include:

- K–5: Investigations in Number, Data, and Space (http://investigations.terc.edu/)

- K–5: ThinkMath! (www2.edc.org/thinkmath/faq.htm)

- Middle school: Connected Mathematics Project (http://connectedmath.msu.edu/)

- High school: CME Project (www2.edc.org/cmeproject/index.shtml)

For additional information about standards-based curriculum visit the K–12 Mathematics Curriculum Center website at www2.edc.org/mcc/. It offers detailed information about all of the National Science Foundation–funded mathematics curricula, as well as an annotated bibliography of research and readings relevant to standards-based curriculum and instruction.

▌ Problems and Lessons

In addition to the resources already listed, teams searching for challenging problems might consider the following sources:

- The Math Forum (www.mathforum.org/) is a comprehensive resource for mathematics teachers and for lesson study teams, a rich source of problems and lessons by topic, and mathematics links. The Ask Dr. Math feature provides information on specific topics.

- The Mathematical Association of America (MAA) (www.maa.org) publications (*Math Horizons*, for example) offers challenging problems that might intrigue teams who are interested in problem solving within their team or looking for problems for their advanced students.

- NCTM journals always include interesting problems. At the NCTM website, there is a searchable problem database, that allows you to access these problems (www.nctm.org/resources/content.aspx?id=16387). NCTM also offers a searchable lesson database, including their *Illuminations* lessons (http://illuminations.nctm.org/).

- Problems with a Point (www2.edc.org/MathProblems/). This site offers a searchable database of engaging problem sets that are always targeted at important concepts, but never routine. An excellent resource for a problem-centered (or problem-solving) lesson.

- Making Mathematics (www2.edc.org/MakingMath). This site was developed to provide problems and other resources for young mathematicians. The problem settings are appropriate for middle and high school students to gain an experience with mathematical research or exploration. Teacher support materials include essays on problem posing and teaching through problem solving.

- Macalester College Problem of the Week (http://mathforum.org/wagon/) offers problems accessible for first-year college students that many teams and high school students would find interesting.

▌ Research Themes

As a team pursues a research theme (like algebraic thinking) over one or more cycles, they will want to do some extra reading about the theme, or even plan to attend a series of workshops on the theme to enrich their understanding. For any given theme or mathematical topic many resources are available. Several strong examples of this type of resource are suggested here.

- The annual *NCTM Yearbooks*. Each yearbook includes multiple essays that treat one teaching issue in depth (measurement, representation, fractions, etc.). Lesson and problem ideas are in every essay. Available at http://www.nctm .org/catalog/.

- *Research-based Toolkits for Mathematics Lesson Study*, Catherine Lewis and Rebecca Perry. 2010. Toolkits provide extensive resources drawn from research (lesson videos, mathematical tasks, student work, and research articles) to enrich team learning. Each toolkit focuses on one mathematical topic that is problematic for U.S. students (area of polygons, proportional reasoning, and fractions). For information on obtaining toolkits see www.lessonresearch.net.

- Cuoco, A., and K. Levasseur. 2003. *Mathematical Habits of Mind.* In *Teaching Mathematics Through Problem Solving*, Schoen, H. editor, Reston VA: NCTM. This article would serve teams considering how to shift from teaching about *topics* to teaching students to *think mathematically*. See also Cuoco, A, E. P. Goldenberg, and J. Mark. 1997. "Habits of Mind: An Organizing Principle for Mathematics Curriculum." *Journal of Mathematical Behavior* 15(4): 375–402.

- Driscoll, M. 2003. *Fostering Algebraic Thinking, A Guide for Teachers, Grades 6–10*, Portsmouth, NH: Heinemann.

- Driscoll, M., R. Wing DiMattteo, J. Nikula, and M. Egan. 2007. *Fostering Geometric Thinking: A Guide for Teachers, Grades 5–10*. Portsmouth, NH: Heinemann.

In these last two books listed above, readings, mathematical tasks, student work samples, and other resources available also in the form of a twenty-hour professional development series could launch a series of lesson study cycles on the themes of algebraic and geometric thinking.

- Stein, M. K., M. S. Smith, M. A. Henningsen, and E. S. Silver. 2000. *Implementing Standards-Based Mathematics Instruction: A Casebook for Professional Development*, New York: Teachers College Press. Readings and activities in this book might provide a framework for teams focusing on increasing the cognitive demand of their lessons.

▌ Readings from Research

Because lesson study provides a rich setting for investigating how students think and learn mathematics, readings with a focus on connecting research to practice are excellent resources for teams.

- Ball, D. L., H. C. Hill, and H. Bass. 2005. "Knowing Mathematics for Teaching." *American Educator* (Fall). The authors' research finds that the knowledge used in teaching mathematics is a complex mix including, for example, common misconceptions, alternate algorithms, representations and how students think about them. These are all topics that lesson study teams delve into deeply.

- Hiebert, J., R. Gallimore, and J. Stigler. 2002. "A Knowledge Base for the Teaching Profession. What Would It Look Like and How Can We Get One?" *Educational Researcher* 31(5): 3–15.

- Kilpatrick, J., J. Swafford, B. Findell, eds. 2001. *Adding It Up* Washington, DC: Mathematics Learning Study Committee, National Research Council, National Academies Press. This resource provides summaries of research on how K–8 students learn mathematics and recommends changes in teaching and curriculum to improve student learning.

- Keeley, P. and C. Rose. 2006. *Mathematics Curriculum Topic Study*. Thousand Oaks, CA: Corwin Press. This book provides guidance and resources for connecting research and standards to topics in the curriculum.

- Lester, F., ed. 2007. *Second Handbook of Research on Mathematics Teaching and Learning.* Reston, VA: National Council of Teachers of Mathematics. A great resource for research on mathematics teaching.

- Ma, L. 1999. *Knowing and Teaching Elementary Mathematics: Teachers' Understanding of Fundamental Mathematics in China and the United States.* Mahwah, NJ: Lawrence Erlbaum Associates. In this book, Ma describes the kinds of knowledge of mathematics teachers have. The concept of *knowledge packages* presented here relates closely to teachers "topic research" in lesson study.

References

Bass. L. E., B. R. Hall, A. Johnson, D. F. Wood, and S. W. Bess. 2001. *Geometry: Tools for a Changing World.* Upper Saddle River, NJ: Prentice Hall.

Carroll, C. and J. Mumme. 2007. *Learning to Lead Mathematics Professional Development.* Thousand Oaks, CA: Corwin Press.

Chokshi, S. and C. Fernandez. 2004. "Challenges to Importing Japanese Lesson Study: Concerns, Misconceptions, and Nuances." *Phi Delta Kappan* 85(7): 520–25.

Driscoll, M. 1999. *Fostering Algebraic Thinking: A Guide for Teachers, Grades 6–10.* Portsmouth, NH: Heinemann.

Fernandez, C., J. Cannon, and S. Chokshi. 2003. "A U.S.-Japan Lesson Study Collaborative Reveals Critical Lenses for Examining Practice." *Teaching and Teacher Education* 19: 171–85.

Fernandez, C. and M. Yoshida. 2004. *Lesson Study: A Japanese Approach to Improving Mathematics Teaching and Learning.* Mahwah, NJ: Lawrence Erlbaum Associates, Inc.

Garmston, R. and B. Wellman. 1999. *The Adaptive School: A Sourcebook for Developing Collaborative Groups.* Norwood, MA: Christopher Gordon.

Gorman, J., J. Mark, and J. Nikula. 2010. *Lesson Study in Practice: A Mathematics Staff Development Course.* Portsmouth, NH: Heinemann.

Karp, J. 2004. "Evaluation Report, Lesson Study Communities in Secondary Mathematics, Program and Evaluation Research Group (PERG)." Lesley University, Cambridge, MA.

Karp, J. 2005. "Final Evaluation Report, Lesson Study Communities in Secondary Mathematics, Program and Evaluation Research Group (PERG)." Lesley University, Cambridge, MA.

Lappan, G., J. Fey, W. Fitzgerald, S. Friel, and E. Phillip. 2006. *Connected Mathematics.* Upper Saddle River, NJ: Pearson Prentice Hall.

Lewis, C. 2002a. *Lesson Study: A Handbook for Teacher-led Improvement of Instruction.* Philadelphia: Research for Better Schools, Inc.

Lewis, C. 2002b. "What Are the Essential Elements of Lesson Study?" *The CSP Connection* 2(6): 1, 4.

Lewis, C. 2005. Video: *How Many Seat: Excerpts from a Lesson Study Cycle.* Oakland: Mills College Lesson Study Group. www.lessonresearch.net.

Lewis, C. and R. Perry. In development. "A Resource Guide for Lesson Study on Developing Number Sense for Fractions." Oakland, CA: Mills College Lesson Study Group. www.lessonresearch.net.

Lewis, C., R. Perry, and J. Hurd. 2004. "A Deeper Look at Lesson Study." *Educational Leadership* 61(5): 18–23.

Lewis, C., R. Perry, J. Hurd, and M. P. O'Connell. 2006. "Lesson Study Comes of Age in North America." *Phi Delta Kappan* 88(4): 273–81.

Lewis, C. and I. Tsuchida. 1997. "Planned Educational Change in Japan: The Shift to Student-centered Elementary Science." *Journal of Education Policy* 12(5).

Liptak, L. 2002. "It's a Matter of Time." *RBS Currents 5(2)*: 6–7.

Lord, B. 1994. "Teachers' Professional Development: Critical Colleagueship and the Role of Professional Communities." In *The Future of Education:Perspectives on National Standards in Education*, edited by N. Cobb, 175–204. New York: College Entrance Examination Board.

Maki, M., ed. 1982. *Kyoin Kenshu no Sogoteki Kenkyu [A Comprehensive Study of Teacher Training]*. Tokyo: Gyosei.

National Council of Teachers of Mathematics. 2000. *Principles and Standards for School Mathematics*. Reston, VA: Author.

National Council of Teachers of Mathematics. 2006. *Curriculum Focal Points for Prekindergarten Through Grade 8 Mathematics: A Quest for Coherence*. Reston, VA: Author.

Perry, R. and C. Lewis. 2008. "What is Successful Adaptation of Lesson Study?" *Journal of Educational Change, 9*: DOI 10.1007/s10833-008-9069-7.

Shimizu, Y. 1999. "Aspects of Mathematics Teacher Education in Japan." *Journal of Mathematics Teacher Education* 2: 107–16.

Stein, M. K., M. S. Smith, M. A. Henningsen, and E. A. Silver. 2000. *Implementing Standards-Based Mathematics Instruction: A Casebook for Professional Development*. New York: Teachers College Press.

Stepanek, J., M. Leong, and R. Barton. 2008. "Improving Mathematics Through Lesson Study." *Principal's Research Review* 3(6): 1–7.

Stigler, J. W. and J. Hiebert. 1999. *The Teaching Gap: Best Ideas from the World's Teachers for Improving Education in the Classroom*. New York: The Free Press.

Stimpson, V. 2007. *Evaluation Report of Field Test for Secondary Lenses on Learning: Leadership for Mathematics Education in Middle and High Schools*. Abstract. Newton, MA: Education Development Center, Inc.

Sugiyama, Y. Unpublished paper. Improving Mathematics Teaching and Learning Through Lesson Study, May 20, 2005, presented at Chicago Lesson Study Group Conference. Wasada University: Lesson Study in Japanese Mathematics Education. Translated by Tad Watanabe.

Wang-Iverson, P. and M. Yoshida, eds. 2005. *Building Our Understanding of Lesson Study*. Philadelphia, PA: Research for Better Schools.

Watanabe, T. 2005. "Knowledgeable Others: What Are Your Roles and How Do You Become More Effective?" An Invitation to Lesson Study: A Facilitator's Guide, Handout 13.1 Translating Lesson Study for a U.S. Context. Lesson Study Leaders Symposium, Learning Point Associates, Northwest Regional Educational Laboratory.

Watanabe, T. 2007. Kyouzai Kenkyuu. Presentation at the Chicago Lesson Study Group Annual Conference, Chicago, IL.

Watanabe, T. and P. Wang–Iverson. 2002. Role of Knowledgeable Others. Presentation at the Lesson Study Conference, Stamford, CT, November 20–22. See www.rbs.org/lesson_study/conference/2002/papers/watanabe.php.